EUROPEAN AIRPORT RETAILING

European Airport Retailing: Growth Strategies for the New Millennium

Paul Freathy

and

Frank O'Connell

MACMILLAN
Business

First published 1998 by
MACMILLAN PRESS LTD
Houndmills, Basingstoke, Hampshire RG21 6XS
and London
Companies and representatives throughout the world

ISBN 0-333-69084-2 ✓

A catalogue record for this book is available from the British Library.

This book is printed on paper suitable for recycling and made from
fully managed and sustained forest sources.

10 9 8 7 6 5 4 3 2 1
07 06 05 04 03 02 01 00 99 98

Printed in Great Britain by
Creative Print and Design (Wales), Ebbw Vale

To Iris and Iestyn – greater gratitude than can be expressed in words

To my father, Frank R. for the love of reading

Contents

Preface and Acknowledgements

They say that opposites attract. To me, airports and aeroplanes were always a childhood fascination, now, 25 years later, little has been done to change my opinion or to dampen that early enthusiasm. Airports represent a microcosm of cultural and social diversity. An opportunity for the academic to study, to measure, to test, to hypothesise, to evaluate and, when the last piece of data has been analysed, the opportunity to experience the excitement of flying. The unbridled power that is exuded from a 747 as it begins to take off, the feeling of disbelief as the 400-tonne aircraft struggles into the air and, ultimately, when the drinks trolley is trundled down the aisle, the chance for one to fantasise about the glamour of being an airline pilot.

Fortunately for this book, the publishers and my career, my colleague and co-author Frank O'Connell has a more realistic and sane view of the air transport industry. Honed out of years of practical experience, he has on more than one occasion had to remove words and phrases such as 'dead brilliant', 'fab' and 'Marxist dialectic'. Through his vast array of contacts this book has become a reality. Scribbled notes have become sentences, sentences have become paragraphs, paragraphs have become chapters and ultimately chapters have become drafts. Collaborative work between academia and industry has much to commend it and this partnership has been one that we have both enjoyed.

Our objective in undertaking this research was to compose a text that had both an academic foundation as well as a practical intent. It has been written to appeal to both an academic and practitioner audience and attempts to convey an understanding of a dynamic and transitory industry. In this sense the book aims to provide a practical understanding of the factors that influence, structure and mediate the functioning of airport retailing.

Apart from the roars of self-congratulation and congenial backslapping that took place when this book was finally completed, there are a number of individuals who have provided us with considerable assistance. Help it is said comes in many forms, those who provided us with interviews include Guntram Brendel, Ray La Comber, Doug Newhouse, Theo van Alphen, Joke Lutterman, Tony Haines, Wilco Sweijen, Roger

Coleman, Derek Hughes, Johan Heins, Oliver Costello, Breda Grant, Bruno Lesser and Ann-Marie Brett. Help with gathering technical data and information came from Wendy O'Connor, Alan Leavy and Jean Taylor. Special thanks to Jacques Parson for arranging the Schiphol interviews and providing such a forthright insight into his own company. Our gratitude also extends to John Shanahan, Tom Haughey and Miriam Ryan for taking the time to read the draft chapters and providing many useful comments.

Special thanks to my dear son Iestyn who encouraged a rewrite of Chapter 7 by dropping my original manuscript into our toilet. Ken – thanks for the phone call.

My special thanks to Carmel, Aisling and Fiona for their patience both in Tavnagh and Dublin during my many hours at the keyboard.

P.F.

F.X. O'C. November 1997

List of Tables

List of Figures

1 Airport Retailing in the Context of Airport Development

Introduction

During the 1980s it became a popular battle-cry from many academics and practitioners that there was little research and published work on the retail sector. Indeed it was considered to be a relatively neglected area of study. In the 1990s that claim remains more difficult to justify as anyone prepared to browse through the voluminous retail literature will testify. Yet despite this welcome redress, the retail sector cannot be said to be fully investigated. Partly owing to its dynamic, constantly changing structure and the newly emerging consumer markets that characterise the sector, retailing provides a host of opportunities for in-depth investigation. One of these rapidly emerging markets is airport retailing. Despite the fact that tax/duty-free retailing celebrated its 50th anniversary in 1997, little in the way of published academic research has been undertaken.

Within the UK, airport retailing has become the focus of both academic and practitioner attention. A number of reasons account for this general increase in interest. First, there has been an expansion in ancillary forms of retailing such as in petrol forecourts (Denning and Freathy, 1996) and in hospitals. Airports have been classed under the general heading of proximity retailing and have occupied a secondary focus in the study of retailing. Their increasing prominence may be attributed to the structural change that has occurred within retailing over the past three decades. Bromley and Thomas (1993) described such a change as a revolution. In the UK, for example, Fernie (1995) identifies sequential waves of retail decentralisation out of the city centre. The first two waves led to the development of edge-of-town and town-centre retailing and was championed first by the grocery multiples and later by the DIY, electrical, carpet and furniture retailers. The third wave, identified by Schiller (1986), was in response to the relaxation of planning regulations. It was argued that this would further encourage out-of-town developments and comparison shopping, and fuel the growth of regional shopping centres.

The impact of the third wave never fully materialised, constrained by the 1987 stock-market crash, prolonged recession and government vacillation (Fernie, 1995). More recently a further challenge has impacted upon the retail sector. The growth of out-of-town shopping developments has arguably approached saturation and the implementation of Government planning restrictions has constrained the extent of out-of-town development. The fourth wave of decentralisation has manifested itself in the form of warehouse clubs, factory outlet centres and airport retailing.

Airport retailing is often used as a substitute term for duty-free and the two are used interchangeably. While the importance of the duty-free industry to airports cannot be denied and will be examined in greater detail in later chapters, this book has a wider scope as it also examines tax-free and tax- and duty-paid retailing. Its parameters are however limited mainly to the European context. This focus is due to a number of reasons. First, taking a global view of airport retailing would require at least a second volume of text. Secondly, and perhaps more contentiously, it will be argued that some of the most successful and sophisticated airport retailing can be identified within the European arena. Humphries (1996) maintains that European airports generate 46% of world airport sales. This is not to suggest that the book will exclude other examples where necessary but its primary focus will be upon Europe.

The Air Transport Industry

Before one considers the role that retailing plays in the commercial success of an airport, it is helpful to provide both a contextual and conceptual framework for understanding the industry. Retailing does not happen in a vacuum and will be influenced and structured by changes in the wider environment in which it operates. Air transportation is a dynamic, global industry that has altered radically over the past three decades. Having a thriving international airport or successful national airline can provide a significant contribution to a country's economic prosperity. There remain a variety of advantages that manifest themselves at the local, regional, national as well as international level. The primary benefits that may be expected to accrue include:

- *The generation of economic wealth.* ACI (1992) estimated that the total economic impact on gross world output contributed by the air transport industry amounted to US$1000 billion. The sources of this wealth are many and varied but include: *tourism and tourist-related activities*, since air travel has been successful in opening up previously remote and developing countries and regions; *improvements in economic efficiency*, for example by encouraging faster distribution of products, through enabling Just in Time principles to be formulated, providing regional sales support and allowing the incorporation of foreign sourced components into the production process; *stimulation of entirely new industries*, such as the export of fresh fruits and vegetables from Africa, the Caribbean and South America to Europe.

- *The generation of employment.* If one accounts for direct[1], indirect[2] and induced[3] employment, the aviation industry employs more than 22 million persons, with some estimating this to grow to 30 million by 2010 (ACI, 1992).

- *The generation of tax revenues.* Many countries levy a travel tax on both domestic and international passengers. This is usually added to the price of a ticket, although in some instances it may be collected directly by the airport as a passenger departure tax. Airports and airlines are both generators of significant revenues for governments through local and corporation taxes. It remains common for governments and/or regional and city authorities to be either the sole owner or a major shareholder of its country's airports. Consequently these authorities may share in any dividends generated by the enterprise. In addition, those trading at the airport, such as exporters and freight forwarders, will pay VAT on their sales and services, while social security payments and income tax are generated from the salaries of employees.

- *Locational benefits.* These remain less tangible and quantifiable but their importance for a region or country remains significant. For example, having a local airport can help attract new business to the region by providing quick and direct accessibility to a transportation infrastructure. For example, the Stockley Park development built on an old landfill site next to Heathrow airport has established itself as one of the premier business parks in Europe and has been successful in attracting both international high tech companies as well as

---

[1] This includes persons directly employed within the industry.
[2] This includes persons employed in off-airport activities such as working for airport suppliers, travel agents, hotels and restaurants.
[3] This is calculated through an economic multiplier and relates to the creation of employment based on the activities of those directly and indirectly employed by the industry.

airport-service industries. Similarly, Cologne has expanded its air-freight business after the authorities persuaded the international distributor TNT to use the local airport as a hub. An airport may also be used to help market a region as the centre for industrial or service relocation. For example, the development of freeports or trade zones on the land surrounding an airport provides manufacturers with tax exemptions and incentives on the pre-condition that all the goods produced are for export only. Once established, such companies have the potential to generate significant employment in the area. These zones attract many major international manufacturing corporations, lured to the region by tax benefits and access to international transport facilities.

Air transport has been characterised by a series of significant developments over the past two decades that have led to changes in the industry's structure. Many of these changes have been in reaction to political, social and economic factors outside the control of the industry. Figure 1.1 highlights the main external influences upon the air transport industry.

The demand for air transport has increased steadily over the past two decades and the expectation is that by the year 2000 over 1.5 billion people will travel by air. Globally, passenger growth is expected to

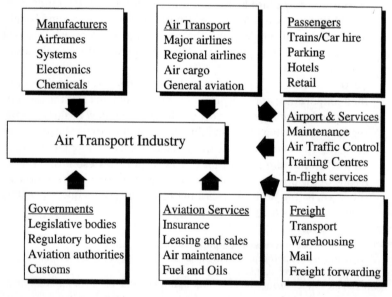

Figure 1.1   *The air transport industry.*

average around 6% per annum with routes to and from Asia being in greatest demand. An ACI survey (1992) predicted that travel in this region would average a growth rate of 8.6% between 1990 and 2010. The rapid increase in air travel as a transport medium has been attributed to a number of both supply and demand side factors. These include:

- *A change in consumer tastes and needs.* Much academic marketing literature highlights how consumers have changed in terms of their preferences, aspirations and outlooks. Individuals are more discerning in terms of the level and quality of service they expect and are increasingly willing to experience foreign travel and cultures. Many countries have been quick to identify the income potential that stems from tourism and have aggressively marketed themselves to the consumer. Aided in part by a wider choice of destination plus the increased leisure time many now enjoy, there has been a growing propensity for tourists to expand their travel horizons. Individuals are looking beyond national boundaries and Southern European charter locations to more exotic long-haul holiday venues in the Far East, Africa and the Americas.

- *A decline in the relative cost of air transport.* The last two decades have seen continuous improvements in aviation technology and efficiency, which, coupled with the liberalisation of the air transport market has led to an intensification of competition. New low-cost operators have entered the market offering alternatives to the traditional state-run and state-controlled airlines. Consequently the relative cost of an airfare has declined as a percentage of an individual's disposable income. An ACI (1992) survey estimated that after accounting for inflation, air fares were approximately 70% cheaper than in 1970.

- *Increased economic activity.* A link has been made between economic prosperity and airline activity. As a country develops economically and its population becomes more affluent there is an increased propensity to use air transport. For example, in 1995 the number of tourists to Vietnam rose by 22.8% (Lloyd-Jones, 1996) and is predicted by IATA to grow by 27.5% per annum up to 1999. Among both developed and developing nations, air travel has become an increasingly important form of travel for the business community. World Trade Agreements have reduced trade barriers between the world's economic zones and increased the propensity and incentive for international business.

- *The growth in international trade.* One of the features of the 1980s and 1990s has been the increased number of businesses looking to expand on an international scale rather than being confined to national boundaries. Retailing in particular has been a sector characterised by low levels of overseas expansion. Competitive pressures, saturation in the home market and technological availability have created the impetus for increased levels of international activity and have assisted in the proliferation of the air transport industry.

Industry Evolution

The air transport industry has undergone major transition over the past three decades and the speed and scale of this change show no indication of slowing. In analysing the evolution of the air transport industry, two theoretical models assist in the process. First, the 'industrial life cycle' (ILC) is a model that although simplistic in its approach, assists in providing a basic conceptualisaton of the evolutionary stages of the development of commercial air transport. Secondly, Porter's (1980) well-cited 'five competitive forces' model represents a means of understanding the current levels of competitiveness within the air transport industry.

Similar in many ways to the product life cycle, the industrial life cycle asserts that demand within an industry will vary over time (Figure 1.2). The *embryonic* stage is where growth is slow and the industry is beginning to develop. The *growth* stage is where new consumers enter the market having become familiar with the product. At this stage there is rapid growth and potentially high margins, with little competition, although the industry is also characterised by a high degree of uncertainty.

In the context of the air passenger transport industry, the embryonic stage existed in the late 1940s and 1950s with the development of scheduled flights across the Atlantic and the development of the commercial jet airline. The origin of aircraft services was however much earlier. For example, a remodelled World War I bomber was used to carry passengers from Hounslow Heath across the English Channel in August 1919. Some years later flying boats were used to cross the Atlantic using Foynes on the Shannon estuary as the first landing point in Europe. Growth continued after the war though limited by the range and payload capabilities of aircraft. It was not until the introduction of the *Boeing 707* in 1958 and the *Douglas DC-8* in 1959 that rapid progress in intercontinental aviation was made.

The duty/tax-free industry also began to develop at this time. Its origins are reputed to stem from the 18th century when gentlemen returning from a Grand Tour of Europe claimed they needed sustenance of

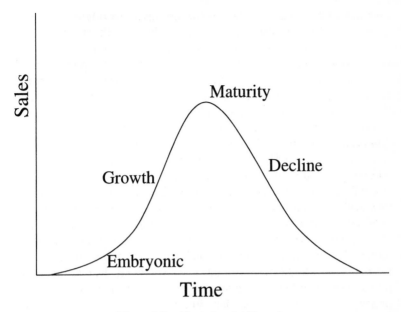

Figure 1.2   *The industrial life cycle.*

spirits and wine on their journey by stagecoach from port to home. It became accepted practice to permit the import of opened bottles of liquor, which later became unopened bottles for the same purpose. In the 20th century the advent of ocean-going liners provided the opportunity for goods which incurred high excise duties at home, to be bought elsewhere at lower prices and consumed on the journey. Thus duty-free sales evolved by both custom and practice.

The Chicago convention of 1944 extended the tax-free status beyond ships in international waters to include aircraft on international flights. It also provided for the creation of 'Customs Free' airports to service these aircraft. In 1947 the Customs Free Airport Act established Shannon in the Republic of Ireland as the world's first duty-free (or 'free zone') airport. Shortly after this the first duty-free shop was opened. This shop was operated under the provisions of Irish Law, which amounted to a waiver of national taxes on goods destined for 'export' beyond Ireland's fiscal jurisdiction. Both embarking and transit passengers were entitled to purchase goods exempt from normal taxes and duty. At this stage however the scale and sophistication of airport retailing remained small, with limited availability and little variation in choice. Initially in the UK, only liquor was permitted for sale. Ambler (1992) notes that when duty free was first available at Heathrow, bottles were carried across the tarmac in a bicycle basket and sold in tents alongside the runway.

While the majority of scheduled flights remained under the control of state operated airlines, the development of the charter holiday industry in the 1960s led to a massive increase in the volume of passenger traffic. Destinations in Southern Europe became the focal attraction for tourists from Scandinavia, Germany, Holland and the UK. Reacting to the growth in these passenger volumes, the duty-free industry also began to grow and became an established part of an airport's operation.

The late 1970s and 1980s represented a period of major change in the air transport industry. A number of charter airlines experienced financial difficulties and Europe witnessed a number of mergers by state owned airlines. More fundamentally, in the early 1980s the deregulation of air travel in the USA represented a catalyst for major structural change within the industry. The development of 'wheel and spoke' and 'hubbing' in the USA established major transfer points through which the majority of flights were routed. The airports involved expanded significantly while others on the periphery experienced lower growth rates or saw traffic diminish. A similar situation was experienced in Europe when a number of airports began to emerge as hubs, drawing an increasing number of the newly created airlines to these locations, and intensifying the pressure upon established airlines to embark upon a price competitive strategy. As many of Europe's established airlines were state owned, their cost structures reflected a regime of regulated air fares and route monopolies. The new entrants to the airline market used the competitive advantage afforded by their lower cost base to offer discount fares as a way of growing market share. Companies such as Laker and Air Europe provided low-service, price-based alternatives and sought to compete directly with established airlines.

In a *mature* market, competition remains intense, growth slows and profits begin to fall. Excessive capacity exists as new companies enter the market and embark upon price-based competition. A saturated market occurs when there remain few first time buyers and the majority of demand is limited to replacement products. Less competitive companies who have previously entered the market will be forced out as the market goes through its *shakeout* period. Eventually the *decline* stage occurs when the industry experiences negative growth. The factors leading to a decline of an industry may be technological (e.g. computers replacing typewriters), social (e.g. the growth of health awareness and the decline of tobacco-related products), or demographic (e.g. the baby boom and the demand for children's products).

The structure of the air transport industry within Europe may be considered mature with established carriers, airport operators and an existing infrastructure. However the recent liberalisation of the industry is leading to significant additional growth. New point-to-point routes are opening using secondary airports and offering the consumer lower fare alternatives. There remains no evidence therefore that the industry is entering the decline phase. Significant growth potential exists as new markets in Asia and South America emerge, new routes open and passenger volumes increase.

While the ILC may be criticised for its simplistic approach, its linear portrayal of historical events and its failure to acknowledge the complexities of many industries, the model continues to remain a useful method of conceptualising the chronological evolution of the air industry. As a basic framework it provides for an understanding of the historical development of air transport. However, owing to the diversity of international regulations, the differing levels of state support and control, combined with alternative cultural and socio-economic power structures, the ILC remains unable to provide either an analysis or indication of the current levels of competitiveness within the industry. The strategies followed by airlines and the airport authorities are multi-causal, i.e., they remain the outcome of a number of competing factors. Porter (1980) put forward a model that attempted to identify these influences and the forces driving industry competition (Figure 1.3).

The strength of competitive forces within an industry can be assessed on the basis of these factors. In the context of the air transport industry, buyers may be seen as the airline operators. The greater availability of hubs especially for long-haul routes has improved the bargaining power of many airlines. Given the high proportion of passengers using hubs for transit purposes it matters little whether the airport used is in Denmark, France or Italy. Airlines have been able to use this situation to their own advantage and have traded the threat of relocation off in favour of better financial terms and conditions. Such developments have put pressure on the airport authorities who need to be increasingly proactive in the way in which they market their facilities and passenger services to the carriers.

The power relationship between buyer and supplier is not wholly asymmetrical and the airports themselves are able to exert a degree of influence upon the airlines. The basis of their power stems from two sources. First, many authorities occupy a monopoly position over a particular destination and control the only airport in the region. For example, many charter destinations such as Grand Canaria and Palma are served by only one airport. So long as the airline continues to fly to that destination

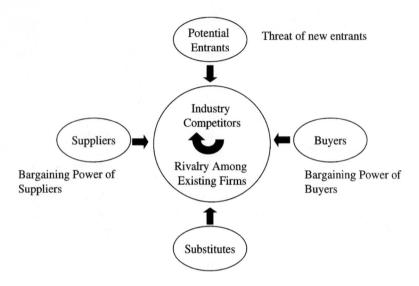

Threat of substitute products or services

Figure 1.3 *The forces driving industry competition (source: Porter, 1980).*

then it is compelled to use the services of a single provider. Secondly, when an airport attains a critical mass of passengers its bargaining power as a supplier increases significantly. As more passengers choose a particular hub, more airlines will want to operate from that airport. This in turn leads to more destinations being available for passengers to choose and creates a lucrative inter-lining market for airlines to access. For example, the large number of people choosing Heathrow as a transfer point has led to airlines having to negotiate with the airport in the hope that they may gain take off and landing slots.

The existence of substitute modes of transport continues to pose a threat to air travel. While the extent of any challenge will depend upon the country in question and the relevant levels of infrastructure development, there remain indications that the existence of alternative forms of travel is impacting upon the industry. Within Europe the creation of fast rail links has had an adverse impact upon some short-haul carrier routes. The TGV link between Paris and Lyon and the Anglo-French *Eurostar* route through the channel tunnel have both had a negative impact upon the corresponding flight routes. With the increasing development of rail links into the main airport hubs this trend is predicted to increase.

Both the entry and exit barriers into and out of a market influence the level of competition within an industry. Traditionally the air transport industry has been a closed industry with bilateral and multi-lateral agreements between countries, regulating the number of flights between destinations. Entry into the scheduled airline market has therefore been restricted and effectively closed to commercial carriers operated by the private sector. Passengers wishing to use air travel have been provided with little choice other than to use the established state-run airlines. The full liberalisation of air travel within the European Union and the movement towards an 'open skies' policy within the USA has led to a number of new airline operators entering the market. While barriers to entry still exist and new entrants will continue to find it difficult to gain slots and gate space, the removal of restrictive regulations will mean an intensification of competition on the part of airports attempting to attract new carriers.

Airport Strategies

The changes described in the first part of this chapter have placed considerable pressures upon the airport industry. Increased liberalisation, an intensification of competition and growing passenger volumes have all helped to create a series of imperatives to which the industry must respond. The reaction by the airports to these demands has varied by country, by region, by size of airport and by individual operator. Consequently it remains difficult to put forward any single generic approach that explains how an airport has coped with the challenges it faces. One way of attempting to conceptualise the pressures placed upon an airport authority is illustrated in Figure 1.4.

Airport authorities will be subject to the demands of numerous private and state-owned interest groups who will seek to exert pressure upon the operation for different, even contradictory reasons. For the airport operator the objective will be to balance the demands of these different parties while at the same time developing a strategy that will guarantee its survival in the long term. The levels of influence exercised by each of these groups and the power of the airport authority itself will structure the strategic direction of the airport. Despite having the same objectives (of moving passengers between one transport mode and another), the approach taken by different airports will often vary significantly.

Traditionally airports have been administered and controlled directly by the state or by a body appointed on its behalf, such as a Government-

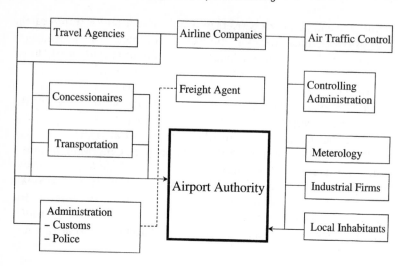

Figure 1.4   *Airport system connections.*

controlled company dedicated to the management of airports. A number of reasons have been put forward to account for the state's involvement in managing airports. Smith (1994), for example, maintains that because many cities have only a single airport, their role becomes central to a country's economic and social development. Keogh (1994) notes that in the Irish Republic, the high degree of state involvement in airport ownership is due to the relatively recent status of Ireland as an independent economy. Control over the country's airports is therefore central in ensuring not only general economic growth but also as a way of assisting in rural redevelopment and regional policy.

While it remains accurate to suggest that the majority of airports throughout the world still have some form of public sector ownership, the level of operational control exercised by central Government may be balanced against a greater participation from private sector interests. A reduction in the control and regulation of the airport industry by the state, in favour of greater commercial sector involvement, can therefore be identified. The Portland Group (1996), for example, identifies four methods by which the private sector may increase its involvement in the management of airports. These are:

Total or partial sale of an airport to the private sector, including long leases: This covers both outright ownership and leases of such long duration that effective control is passed to the operator. There remain a number of methods by which an airport may be privatised. These include:

- flotation, where the share capital is issued and subsequently traded on the stock market;
- the placing of shares with a pre-determined number of financial institutions by the vendor's banking advisors;
- raising finance through loans and bonds;
- using a tender or trade sale to secure a single buyer or conglomerate of buyers to operate the airport.

While currently only the British Airports Authority (BAA) operate a totally privatised group of international airports, the Governments of Austria and Denmark have sought to reduce their interests in Vienna and Copenhagen airports. This may be achieved either through a partial stock-market flotation or as in the case of the recently privatised Australian airports of Melbourne, Brisbane and Perth, a trade sale to a company or conglomerate of companies.

The outright privatisation of an airport however continues to remain uncommon. This is primarily due to the monopoly position many airports hold. Governments may be cautious about the creation of private sector monopolies over which they will exercise little strategic or day-to-day control. Even in the case of the BAA there remain restrictions on its operating licence. For example, the company does not have the freedom to determine its own landing charges, these being set by the Civil Aviation Authority (CAA). In 1992 the CAA provided the BAA with a five-year formula that limited its charges to 8% below the retail price index for two years, 4% below in the third year and 1% below for the last two years. In 1996, a new five year formula was set following a review by the Monopolies and Mergers Commission. This limits charges at Heathrow and Gatwick to 3% below the retail price index and allows charges at Stansted airport to be 1% above the RPI (MMC, 1996). State control has also not been completely devolved. The UK Government continues to hold a single 'golden share' and in the area of aircraft operation activity, retains responsibility for setting safety and security standards (Sewell-Rutter, 1995).

Granting of a time-limited lease or concession to the private sector: This represents a popular method of utilising private sector resources. The responsibility for the airport is given over to an operator for a fixed duration and therefore provides Governments with a greater degree of control than in an outright sale. The types of lease and the financial terms agreed will vary but may involve an initial up-front payment, a guaranteed level of investment or the payment of annual fees.

Private sector management using a management contract: This method of contractual relationship, although not widespread, is becoming increasingly popular. In return for a management fee, the contractor takes responsibility for all elements of the airport's operation. To ensure that the operator maximises the utility of the airport, specific performance-related bonuses are built into the contract. For example, BAA have been contracted to run Indianapolis airport under such an arrangement. As will be illustrated in Chapter 2, management contracts do not have to cover an entire airport, but can be applied specifically to the retail operation.

Private sector construction and operation of terminal buildings: Under this method, an airline has an airport terminal constructed on its behalf that it then leases from the airport operator for a specified period of time. This arrangement provides both a guaranteed revenue stream and allows the airfield and the land to remain under the control of the airport. The airline may get exclusive rights to the terminal or receive substantial discounts on its usage. While popular in North America and Australia, discriminatory cost-related pricing is unlawful in Europe and this method of privatisation is prohibited. Governments are however examining methods by which private-sector organisations may take responsibility for new airport and terminal development. One such method is the BOT (Build, Operate and Transfer) system whereby in return for the right to operate an airport or terminal over a specified period, the concessionaire **builds** and **operates** before ultimately **transferring** the facility back to the statutory owner.

Each of the options described above requires varying levels of control to be devolved to the private sector. While the movement from a state-controlled, centrally planned industry to a privately owned commercial concern is not necessarily the future blueprint for all airports, the evolution of the sector has led it to become less reliant upon government support and more focused upon the generation of revenue. While the factors that have accounted for this change are multi-causal, it is possible to identify a number of key influences that have prompted such developments.

Why Privatise?

The gradual liberalisation of air travel within the European Community led to rapidly increasing air traffic volumes and an intensification of competition. A consequence of this volume growth has been considerable terminal and airfield capacity problems. Airports such as Heathrow maintain that terminal capacity is close to saturation. It has been estimated that up to £600 million will be lost annually by 2005 unless a fifth terminal is built (BAA, undated). Doganis (1995) highlights the problems of runway shortage maintaining that the rapid growth in air travel has been outstripping available capacity. It is argued that by the year 2000 between twenty five and thirty European airports will suffer serious runway congestion.

Airports have therefore embarked upon major capital investment programmes not only to accommodate increased passenger traffic but also to cater for a new generation of aircraft. If the creation of the 800–1000 seat supercarriers becomes a reality, airports will be required to enlarge current terminal capacities and provide specific taxiing, parking and take-off facilities. Such developments further highlight the requirement for large capital investment programmes on the part of the airports.

While the liberalisation of air transport in Europe may be seen as the catalyst to increased competition, airport privatisation could not have occurred without state sanction. This has been forthcoming for a number of reasons. Smith (1994) maintains that deregulation was prompted primarily by the desire of Governments to avoid the financial burdens associated with subsidising airport capital investment. Airports have traditionally had to compete with other areas of public expenditure such as education, health and defence. The increasing pressures associated with operating an airport have arguably led to a realisation that airports are in a mature, competitive market and need to be run on commercial rather than state principles.

This has led some to maintain that the management culture within the public sector was often unsuited to meet the demands of the commercial marketplace (Doganis, 1995; Sewell-Rutter, 1995). A less bureaucratic and formalised structure exists within the private sector and provides management with the financial flexibility to meet the needs of shareholders and provide greater freedom of action.

Allowing private-sector organisations a financial interest in airport operations may therefore be seen as an efficient and cost-effective way for the state to maximise revenue while at the same time improving customer service and quality standards. The level of return is increased while the degree of risk is minimised as the state draws upon a specialised set of management skills.

From an investor's perspective, airports remain a viable business opportunity. The attraction lies in the expectation that the financial returns on the investment will be greater than could be achieved elsewhere. Sewell-Rutter (1995) identifies four factors that account for the high expectations of investors. These are:

- the growth of civil aviation activity combined with the expectation that leisure-based travel will continue to increase;
- the monopolistic location of many major airports;
- the use of major airports by international airlines who are not as susceptible to local or regional economic cycles;
- the availability of real estate.

Airport Commercialisation

The increased involvement of the private sector in the management of airports has been accompanied by the growing awareness of the commercial potential that a well-run airport can offer. While providing a number of opportunities for income generation however, the role and purpose of an airport would seem to be far from defined. There are those who remain close to the traditional view of an airport, i.e. that it exists to ensure the efficient movement of passengers between one destination and another. An alternative, and perhaps more eclectic approach, views airports within the framework of consumer change. In this context airports are seen not only as modal interfaces but also as leisure attractions and primary destinations in their own right. If airports are viewed as locations through which passengers are to be moved as quickly and as efficiently as possible, then the role of commercial activities within airport operations will always remain limited. If however an airport is viewed as a primary leisure destination in itself, then it will remain possible to develop further the commercial opportunities within the airport. The priority assigned to the generation of such revenues will depend upon a number of factors, not least the extent of state support for such a strategy and the level of commercial representation at the most senior levels within an airport's management structure.

Doganis (1992) examined airports in the context of commercial and aeronautical activities and drew a distinction between the *traditional* and *commercial* models of operation. The traditional model views retailing as supplemental to the airport's main activities of ensuring consistent passenger flow, the servicing of aircraft and the maintenance of the airport. Doganis *et al.* (1994) maintained that despite managing and operating forty airports during the early 1990s, the Spanish airport authority

(AENA) fell into this category. Its commercial activities remained relatively undeveloped and consequently it relied more heavily upon aeronautical revenues.

The capital required to develop and maintain airports is generated from two sources: commercial and aeronautical activities. Papagiorcopulo (1994) maintains that non-aviation revenues have increased in importance, primarily because income derived from aeronautical activities has declined as a result of intense competition within the airline industry. The charges levied on airlines by airports, for using their facilities, have remained relatively static since the late 1980s. This has partly been in response to Government policies aimed at encouraging in-bound tourism and travel, and partly because the airline authorities are operating on limited margins and keeping fares low. Nevertheless, aeronautical income remains an essential part of airport economics. Four principal sources of revenue exist, as follows:

Aircraft landing fees: Traditionally, landing fees have represented the major source of revenue for an operator. While some airports levy both take-off and landing fees, the majority of authorities have only an arrival charge. The method of charging varies but is often based upon the Maximum Take Off Weight (MTOW) or the Maximum Authorised Weight (MAW) of the aircraft. Airlines are charged to the nearest tonne, 1000 lb or 500 kg. What is included within the landing fees varies but may include air traffic control charges, security and policing costs, landing and parking facilities, passenger use of the terminal and take-off facilities. Currently the cost to land an aircraft in Europe varies between airports. For example, Table 1.1 illustrates the range of turnaround charges in 1994 for a *Boeing 747* carrying 278 persons and weighing 325 tonnes.

Table 1.1   *Airport fee charges for landing, take-off and passenger arrival for selected European airports in US$ (1994).*

Vienna	8524
Spain (all Class 1 airports)	3598
Italy (all state airports)	4212
Dublin	6712
Schiphol	7215
Glasgow	9955
Newcastle	11415
Heathrow (peak rate)	7335
Birmingham	9955

Sources: various.

Doganis (1993) maintains that landing fees will continue to remain an essential source of income for the airport operator. Unlike the airlines who have had to face increased fuel prices and competitor activity, airports have been able to continue charging landing fees regardless of the number of passengers or the amount of cargo being carried on the aircraft.

One consequence of the liberalisation of the commercial air transport industry has been the increasing pressure on airports to maintain their landing fees and charges at existing rates or even to reduce them below current levels. For example, the basic airport charges levied by Aer Rianta, the Irish airport operator, upon carriers has remained fixed since 1987. In addition a discount scheme has been introduced and made available to all airlines. Traffic generated from new routes incur minimal or zero charges in the early stages of development, while additional traffic on existing routes often receive significant discounts. Low-cost budget operators in particular have campaigned hard to ensure that airport operators do not increase landing charges. Where they have been unsuccessful, the airline will often reschedule the route to an alternative airport.

Aircraft parking and hangar charges: The aircraft landing fee paid by the airline covers the cost of aircraft parking for only a specified period of time (usually two hours). If the aircraft is parked on the apron for longer than this, then it becomes liable for parking surcharges. These tend to be levied on a 24-hour basis and are in the majority of cases related to the tonnage of the aircraft. Hangar parking is also offered at some airports but the demand for such facilities remains considerably lower owing to the relative costs.

Passenger charges: This is a fixed fee charged by the airport in addition to the aircraft landing fee. It is applied to each passenger leaving the airport and is different from the departure tax (which is a Government tax) common in many countries. The majority of travellers are unaware that such a charge exists as it is normally absorbed into the ticket price.

Passenger/Ground handling charges: It is the duty of the airlines to guarantee the safety of their passengers and their accompanying baggage both in the air and during their time in the terminal building. Responsibility extends from the time individuals check in until they leave the destination airport with their baggage. Non-flight activities such as check-in, baggage handling and passenger boarding are known as ground handling. The control of such operations represents a lucrative income stream. Some airport authorities, such as Frankfurt, have decided to take responsibility for all ground-handling and charge the airlines for their services. More traditionally, the ground-handling task has been undertaken by the national carrier. Such an arrangement has brought criticism from other airlines and the European Commission as it effectively creates a monopoly and can in some instances represent a barrier to entry. Alternative ground-handling arrangements include subcontracting to a third party operator under a concession arrangement, or the airlines may self-handle, i.e. undertake ground-handling themselves. A recent review has led to an EU directive which requires larger airports to open their ground-handling operations to competition and provide user airlines with the option of self-handling or the choice of more than one third-party handling company.

In addition, the airport may generate revenue through the provision of support facilities such as apron services, security and airbridge fees and through royalty charges on fuel throughput. Doganis (1995) estimated that airport authorities in Western Europe derived approximately 58% of their income from aeronautical activities. However evidence suggests that maintaining income from such sources is becoming increasingly difficult for airport operators who have had to look at alternative ways of supplementing their revenue streams.

The decline in aeronautical revenue has been compensated by the development of commercial activities within the terminal. The increased number of passengers using an airport provides a number of potential income earning opportunities. The contribution that non-aeronautical revenue makes to total operating profits varies between airports and will be examined in more detail in the next chapter. The main areas of commercial activity however include:

Income from rents and leases: Airports often own significant tracts of land which may be used for development purposes. While the construction of hotels remains commonplace, airports have also established enterprise zones and industrial parks within their perimeters. Large hub airports and those located close to major road and rail networks offer the greatest potential in this respect. Amsterdam's Schiphol airport, for example, has over 1.1 million square metres of freight facilities, business parks and airport buildings and in 1995 derived 13% of its operating income from rents and leases. Rents are charged at market rates as airport properties represent prime sites because of their proximity to the transport infrastructure.

Airline operators are also a significant source of non-aeronautical revenue. They may decide to lease sections of the airport for activities such as cargo and freight movements. This very often includes the renting of hangars and shipment areas. Office space is also leased to airlines and the many other businesses operating within the airport, including car rental firms, service providers and ground-handling companies. In addition, ticket and service desks, business lounges and business centres may all be leased out by the airport operator.

Concessionaire income: Fees from concessions represent a significant proportion of the airport authority's income. The use of concessions is widespread and includes retail outlets, car rental facilities, car valeting, banking and foreign exchange facilities, bar and restaurant services. Those operating a concession range from established, international companies to specialist airport traders. In the majority of instances, concessionaires pay a percentage of their turnover to the airport operator plus a minimum guaranteed sum.

Direct sales income: In some instances the airport authorities may decide to run a number of the commercial activities themselves. For example, the duty/tax-free shopping, in-flight and ground catering and car parking. In the case of retailing, the airport remains responsible for all stock holding, merchandising policies and operational decisions related to the running of the store. While the risk associated with this form of commercial activity is greater than with a concessionaire strategy, all income accrued goes directly to the airport operator.

Other commercial activities: This category covers all other revenue-generating activities including left luggage, advertising, petrol stations and taxis. Revenues derived from car parking in particular can make a significant contribution to profitability. The increase in business travel has contributed to the increased demand for car parking space. For example, 7.5% of Schiphol airport's operating revenue in 1995 was derived from car parking fees. In the majority of cases, car parking is sub-contracted out to a third party operator although in some instances it is operated by the airport authorities.

Against these income streams the airport has to balance a number of different costs. One of the most significant financial undertakings for an airport will be its capital development programme. New terminal buildings or the construction of an additional runway require substantial injections of capital. Traditionally this has come from the public exchequer, though more recently finance has been raised from a combination of the airport operator's own resources and borrowings. Charges made for these new infrastructural developments relate to depreciation on the capital and interest paid on borrowings. Such charges can represent between 20% and 35% of total costs.

Perhaps the largest single expense for an airport however is its employee costs. Doganis (1992) maintains that staff represents the largest single outlay for an airport (42% of total costs). Staff activities may include passenger/baggage handling, freight handling, catering and retailing. Staff costs will vary depending upon the level of commercial and aeronautical activity undertaken by the airport. For example, if an airport operator decides to run both the retail and baggage operation rather than use concessions, then its staff costs may be as high as 65% of total costs. In addition to the expenses outlined above, an airport's other main costs will include services (water/electricity etc.), maintenance and repairs, security and administration. Table 1.2 highlights the major running costs for BAA during 1996.

Table 1.2   *Running costs for BAA plc 1996 (%).*

Cost	%
Staff costs	28.1
Retail	19.6
Property	16.6
Maintenance	7.5
Police	4.6
General expenses	7.3
Depreciation	11.8
Net Interest	4.2

Source: NatWest Securities (1996).

Threats to the Development of a Commercial Strategy

Capitalising upon the growth in passenger traffic, a number of airport authorities have been highly successful in developing their commercial operations. A significant part of their strategy has focused upon developing traditional duty/tax-free retailing to include a broad range of merchandise tailored to the lifestyles of the various passenger groups. The products remain positioned in the tax-free environment and continue to offer the traditional price differential relative to the domestic market. Such an approach has on occasions provoked criticism from both consumer associations and retailers. Some consumer groups maintain that despite being cheaper than the high street, duty/tax-free products remain over-priced. A number of retailers have argued that duty/tax-free retailing represents a form of unfair competition. However the expansion of airport retailing to include a wide range of merchandise categories has meant that many high street retailers are now trading successfully in Europe's airports.

The impact of this criticism in terms of practical effect has remained limited. What now represents a much more significant threat to airport operations and activities are the legislative provisions currently in place or proposed by the EU. Five key proposals have the potential not only to create a series of difficulties for the airport industry, but also to impact upon their income-earning activities. The main legislative provisions include:

- no frontier/immigration controls for intra-community travellers;
- no customs barriers for intra-community travellers and traders;
- transparency in airport accounting with no cross-subsidisation;
- community wide procurement policies;
- competition legislation affecting aviation and airport commercial activities.

Such proposals will have a number of implications for the air industry. Doganis (1995), for example, notes that intra-community traffic previously considered to be international would have to be reclassified and, in theory at least, handled through each airport's domestic channel. Doganis maintains that at airports such as Gatwick, domestic traffic would increase from 6% to 60%. Such a reclassification it is argued will put domestic facilities under considerable strain and require significant capital expenditure in order to cope with increased passenger numbers. Creating a third stream of international passengers, not liable to customs or immigration controls, has also been ruled out as an option. Apart from its cost and the need to replicate many of the service facilities, the design of many airports would make such a strategy unworkable.

For those airports who have signed the Schengen agreement (which ends all intra-member frontier barriers), passenger reclassification has already become a reality. The agreement has required airports to divide their terminals into Schengen and non-Schengen areas and separate their passengers accordingly. It has also involved significant capital investment in order to duplicate terminal facilities with a consequent doubling of operating costs. The confusion generated by these new arrangements has led to a serious decline in the revenues from the Schengen only passenger area. Schengen passengers have been uncertain as to whether they are still entitled to purchase duty/tax-free products. In Schiphol and Paris airports, the impact of the Schengen arrangements has led to a decline of up to 40% in some merchandise categories.

However, for European airports the planned abolition of intra-EU duty and tax-free shopping on 1 July 1999 represents the most serious threat. The original intention was to halt all such sales from 1 January 1993, but the EU member states, following a campaign by all sectors of the air, sea and transport industries, agreed a transition period of six and a half years until 30 June 1999. The rationale for the abolition is that since the movement of goods between member states will no longer be treated as 'exports' or 'imports' for tax purposes (in a completed single market), it would be inappropriate to waive the tax and duty on purchases that are currently exempt (Netherlands Economic Institute, 1989).

The removal of intra-EU passengers' rights to purchase duty-free products would have a number of effects. It is suggested that duty-free retailing can act as an important shop window for EU goods and stimulate interest in a nation's leading branded products. Such exposure is argued to be an important stage in the process of product internationalisation and provides a fast and cost-effective method of exposing EU products to new markets. Duty/tax-free retailing is maintained as a separate and discrete distribution channel by customs authorities around the world. This is primarily to prevent any leakage of goods into the domestic market which would result in a loss of Government tax revenues. Such a distribution channel, it is argued, enables EU manufacturers to gain direct access to world markets and to achieve a more immediate distribution of their products.

However the primary concerns in the industry remain social and economic. Research indicates that if intra-EU duty/tax-free sales were abolished, there would be a negative impact upon airlines, manufacturers, airports and ferries. The intra-EU duty/tax-free industry supports 140,000 jobs and in 1995 had sales of $4.8 billion (ETRF, 1996). Total job losses, including direct, indirect and induced, could exceed 100,000, a high proportion being women working on the sales floor.

Airport aeronautical charges on intra-EU air services could rise by as much as 69% should abolition occur and for some airports the revenue-to-expenses ratio could decline from 1.36 to between 1.09 and 1.13. This would have the impact of turning some airports into loss-making ventures (Cranfield, 1997). Charter airlines rely heavily on in-flight duty free. If intra-EU duty/tax-free sales are abolished, it is estimated that these airlines will suffer a net loss of $480 million in revenue and over a thousand jobs would be removed from the industry. Other implications would include an estimated increase in the cost of a flight by between 2% and 6.3%. This could lead to a decline in the demand for flights, with the underlying growth being reduced by 3% per annum and a switching to non-EU destinations. It is estimated that such changes will have a particularly adverse effect upon southern Mediterranean tourism and will lead to a reduction of over half a million charter travellers to Greece (PA Consulting, 1996).

One of the principal benefits of the EU air transport liberalisation programme will be an emerging low-cost scheduled airlines sector. Symons, Travers, Morgan (1997) estimate that the growth evident in this sector would be impeded by the loss of income from on-board duty/ tax-free sales and increased charges. They predict a reduction of over 4 million travellers, a decline in passenger revenue by £10 per person, air fare increases of up to 22% and the elimination of almost three thousand jobs in the sector.

Suppliers of alcohol, tobacco and perfume products would also be adversely affected by the removal of intra-EU duty/tax-free sales. Manufacturers of premium whiskies and cognacs are expected to experience a significant decline in their overall sales as buyers switch to cheaper brands or forgo purchasing altogether. While some duty-paid substitution would take place, the total loss incurred would be between 55% and 60% of sales of core duty-free products at EU airports (IRS, 1997).

However it is the airports that are likely to experience the greatest decline in gross profitability with airside retail sales declining by up to 81% (IRS, 1997). Many airports would be unable to make a profit if duty free were abolished. Gray (1994) highlights the importance of such revenues for European airports. In the USA it plays a relatively insignificant role (Los Angeles 12%; Seattle 11% and San Francisco 5% of total revenue), however Schiphol derives 34% of its revenue from duty/tax-free sales, while at BAA it accounts for 33%. The abolition of duty free will cost BAA in excess of £50 million per annum while in Ireland, Cork airport derives 91.8% of its operating profit from duty/tax-free products (O'Connell, 1993). Regional airports with a high percentage of intra-EU traffic are most at risk and, because of economies of scale and insufficient passenger volumes, are unlikely to find profitable alternative retailing activities.

Strategic Options for Airport Retailers

A number of European airports find themselves in the unenviable position of requiring significant capital investment to meet the growth of new traffic while at the same time facing the removal of one of their chief sources of revenue. Airports have a number of options for dealing with this proposed removal. One obvious way is to increase the level of aeronautical charges levied on the airlines. Apart from the hostile reaction such a move would provoke from the airlines themselves, such a strategy could radically alter the cost structure of the industry.

Doganis (1995) estimates that increases of between 15% and 33% would be necessary to compensate for the shortfall in duty/tax-free revenue, depending upon whether the charges were applied to all airport users or only those operating intra-EU routes. It is argued that such a strategy would not only undermine the European Union's desire for cheap intra-community transportation, but would also positively discriminate in favour of non-member states who remain unaffected by the proposed abolition (IDFC, 1989). Nevertheless the seriousness with which the abolition is viewed has been illustrated in the CAA's recent ruling, which granted the BAA permission to increase its landing charges by 15% per annum for two years if duty/tax-free sales are abolished.

While continuing to campaign to retain the right to sell duty-free products on intra-EU flights, airport operators have been compelled to act proactively to its threatened abolition. The highly competitive nature of the airline industry and declining public sector involvement compound the threat of declining financial revenues and highlight the need for a planned strategic campaign. This has manifested itself in a number of ways and has involved both the continued development of existing services and facilities as well as movement into new commercial ventures.

One strategy has been for airports to reinforce their position as the first choice for both airlines and travellers. This is achieved by emphasising quality and the efficient ways that passengers can be processed in a relaxed, congestion-free terminal. For example, because of the anticipated increase in passenger volumes into the next century and the long lead times associated with airport development, the planning process for a new Terminal 5 at London Heathrow is well underway. If successful, the new terminal will accommodate up to 30 million new passengers per annum and reinforce the airport as one of the major hubs for air travel.

A second development designed to combat the threatened abolition of duty free has been to expand other commercial activities. This has resulted in a marked increase in the amount of space now dedicated to specialist shopping facilities within many of the larger airports. BAA, for example, increased retail floor space (including catering) from 400,000 square feet in 1991 to 575,000 square feet in 1993/1994. Weber (1995) estimates that the potential capacity for BAA airports will be one million square feet by the end of this century. If duty/tax-free sales are eventually removed, the existence of such a diverse range of shopping facilities should represent a credible alternative for airport customers.

In an attempt to further reinforce the commercial aspects of the airport, a series of other initiatives has been launched by management. These have included twenty four hour trading, direct mail and catalogue shopping, and the development of consumer loyalty schemes. More fundamentally, airports such as Schiphol International have taken the process of commercialisation further by developing landside shopping centres aimed at the non-travelling public.

The development of the core business has remained the priority of many airports. There have however been a variety of other initiatives designed to further the interests of the airport operators. One of these has been to embark upon joint ventures with foreign partners. While not particularly new, such a strategy has become increasingly widespread in recent years, prompted by the greater involvement of the private sector in the management of airports as well as the threatened abolition of duty free within Europe. Aer Rianta, for example, operates the three main airports in Ireland, has a major shareholding in Birmingham and Dusseldorf airports, has retail interests in Russia, Ukraine, Bahrain, Beirut, Kuwait, Hong Kong, Cyprus and Beijing, and runs the duty-free outlets at both entrances to the Eurotunnel (Table 1.3).

Some airport authorities such as the BAA have also sought to develop interests outside the airport. After being privatised in 1987, the BAA began to diversify its activities, by developing hotels on its own property and opening a hotel in Belgium. In addition it has agreed to jointly build and operate a high-speed rail link between London and Heathrow airport. The company has also invested in property, manages shopping facilities within a hospital and has bought a freight forwarding company. It has entered into a joint venture agreement with the US property developer McArthur/Glenn to establish and operate a series of designer outlet centres. The first in the UK was a 200,000 square foot, sixty unit operation in Cheshire, followed by a 155,000 square foot, forty outlet centre in Troyes, France.

The skills that operators have developed in managing their airports has a commercial value and a number of authorities have established international consultancy companies through which they sell their expertise. ADP, the Paris airport operator, for example, assisted in the redesign and refurbishment of Cyprus airport, while the Frankfurt authority is developing a marketing strategy for the new airport at Spata in Greece.

Table 1.3 *Aer Rianta and Aer Rianta International (ARI) joint venture partnerships.*

Company	Main operations	Shareholding and comments
Aerofirst Moscow	Moscow Sheremetyevo	33.3% equal partnership with Aeroflot Russia and airport authority
Kievrianta	Kiev Airport	49% (in partnership with airport authority)
Lenrianta	St Petersburg airport and in-flight	48.3% (in partnership with Aeroflot St Petersburg)
Sitop	Russian/Finnish border shops	48.5% (in partnership with Vyborg Regional Consumer Society)
ARI Pakistan	Karachi airport	ARI sold its majority holding and operates on a management contract*
ARI and Phonecia Trading Afro-Asia joint venture	Beirut airport	Joint venture covering a 15 year duty-free and duty-paid management contract, likely to start end of 1998
ARI East Asia	Consultancy services for China National Duty Free Merchandise Corp.	65% shareholding
ARI (Middle East)	Bahrain airport	ARI holds 50% of ARI (Middle East) in partnership with local investors and operates a management contract
ARI Duty Free UK	Eurotunnel outlets	ARI operates the Eurotunnel on a management contract

* A definition of a management contract is provided in Chapter 2.
Source: Duty Free Data Base and Directory (1996/97)

The European air transport industry is a dynamic and evolving industry that has been characterised by a series of significant developments over the past three decades. The legislative and operating environments have been altered radically by the full liberalisation of the market on the one hand and a major restructuring of the principal transport providers on the other. New entrants have helped stimulate a resurgence of growth in the marketplace and provided new opportunities for the airports. In addition to being privatised, airports are also being commercialised on a significant scale across Europe. Retailing has come to represent a principal component in achieving this change. Airport retailing is therefore evolving to become a distinct sector of retailing in its own right, with its own unique characteristics and imperatives.

The planned abolition of intra-EU duty/tax-free shopping poses a serious threat to the prosperity of European airports and, if implemented, will place in doubt the economic viability and future profitability of many airports. The proposal will also have a negative effect on employment, in particular in many of the more peripheral regions of the European Union. Duty/tax-free retailers, airport authorities, airlines, ferries and manufacturers are all resisting abolition of this thriving industry and mounting a concerted lobbying campaign. Airports are attempting to counter the negative impact of abolition through a variety of activities, involving alternative retail formats, joint ventures, collaborative partnerships and diversification.

2 Airports and the Retail Industry

Introduction

In Chapter 1, an examination of how airports functioned provided the context for understanding the contribution retailing can make to an airport's operation. In this chapter the aim will be to highlight the increasing importance of commercial activities and to examine how retailing operates within an airport environment. Airport retailing is different. Just how different depends if you are a passenger or a visitor to the airport, whether you are on an international or domestic flight, or whether you are travelling on business or for pleasure.

Unlike other forms of shopping, the primary objective for the majority of visitors to an airport is access to a mode of transportation rather than the purchase of merchandise. In contrast to the high street, many potential customers will be unfamiliar with the environment within which they find themselves and consequently be in a heightened state of anxiety. They may be unsure of the time available to shop, customs regulations relating to duty-free allowances and how to return the goods once purchased. Duty/tax-free retailers also have to contend with long (often 24 hour) shift working, rigid customs controls and the continued pressure of ensuring product availability. Such differences highlight the unique nature of airport retailing and reinforce the need for airport-specific retail strategies.

Prior to examining these issues in detail, it is useful to locate such developments within the wider context of contemporary retail change. Airport retailing obviously does not occur within a vacuum and is a response to both cultural imperatives as well as competitive pressures.

Contemporary Retail Developments

One of the difficulties associated with any European-wide text is its generalised nature. Exceptions to any rule can always be shown to be flourishing in a context completely opposite to that described. An overview of contemporary retail developments, while providing the framework for an understanding of airport retailing, must therefore be viewed in the broadest terms.

This book will argue that airport retailing is a dynamic and complex sector which the speed of change and the intensity of competition have transformed beyond all recognition over the past two decades. Moreover this transition is not complete and one may expect continued growth and structural readjustment well into the next century. The foundation for this assertion not only lies within the airport industry itself but also upon an understanding that general commercial trends manifest themselves in all retail sectors.

The factors accounting for contemporary retail developments have been numerous. One of the most important however has been change in consumer demand. Individuals have become more discerning, better educated and more widely travelled. This has had an impact upon the types of goods offered, the service provision expected and ultimately the structure of retailing itself. Alexander (1997), for example, notes that the key values of many consumers now include quality, value and choice. From a retailer's perspective this places increased emphasis upon developing a retail environment that meets these key criteria.

Household structures have also changed, people are now living longer and the average age of the population in many European countries is increasing. Moreover consumption trends reflect increased female participation in the workforce and a growth of dual income families. Time has become a critical factor, since consumers require both convenience and a faster, more efficient service. In response, stores have been located on frequently used travel routes and formats have been redesigned to provide speed of delivery without queues (Dawson, 1995). More fundamentally, these demographic and socio-cultural changes have led to a growth in the demand for leisure-based activities. Increasingly, families are attempting to draw a balance between work and leisure. The requirement for quality time has prompted the development of a wide range of leisure services and activities from theme parks to holiday clubs to adventure treks in exotic locations.

Changes in consumer behaviour and household structure have had a number of implications for the retail sector. While the total number of small firms within Europe has declined as a whole, there has also been a polarisation of operating scale. In order to satisfy the increased demands of consumers there has been a growth in the number of specialist retailers. Specialists offer a limited number of product categories but provide greater range and depth than previously available in the traditional department store. Companies such as The Body Shop, Sunglass Hut and

Bally epitomise the small unit specialist trading ethos. At the other end of the operating scale there continues to be a growth of large, one-stop shopping outlets. The hypermarkets in France and Germany and the superstore concept in the UK offer convenient access to a wide array of household goods and services.

The rise in the value and volume of retail sales illustrated by Dawson (1995) has been accompanied by an increasing concentration of market share among a relatively small number of firms. This concentration has occurred at the expense of small operators who have seen their numbers decline significantly over the past three decades. In the UK, for example, between 1961 and 1971 the number of independent retailers declined by 12.5%, while between 1971 and 1994 this figure fell by a further 56% (SD10, 1975; SDA25, 1996). The catchment area for those stores that continue to trade is consequently expanding. In 1955 there were approximately 72 persons per shop in the EEC, by 1991 this had risen to 105, and by 1996 the figure was 140 (Tordjman, 1994; Corporate Intelligence Group, 1997).

The growth of large multiple retailers has had a significant effect upon relationships within the supply chain and has led to a reassessment of the power balance between channel members. The product volumes now demanded by the retailer have provided them with a greater degree of control over both the channel and its management. As a reaction to competitor and consumer demands, retailers have initiated a cultural change in the supply chain by imbuing a degree of responsibility upon all members. Terms such as 'partnerships' and 'co-operation' signify a philosophy of mutuality rather than conflict, with an emphasis upon long-term rather than short-term benefit.

The domination of multiple retailers in a number of different retail sectors has provided a number of internal control issues. Increasingly, retailers have looked to derive economies of scale from their operations whenever possible. The outcome of this has been an increasing separation of conception from execution; that is, centralisation of decision making at head office with stores following procedural rules and regulations. Functions such as buying, merchandising, distribution and pricing are all controlled centrally. Stores are viewed as profit centres with an operational rather than strategic remit. While this has led to a number of motivational issues for managers at the store level, it can be argued that the benefits accruing from centralisation have far outweighed the disadvantages (Freathy and Sparks, 1995).

The investment in and utilisation of new technology has been a central factor in understanding the success of individual retailers. There remain few operational areas where technology has not provided a significant contribution to organisational efficiency. Scanning and stock control systems, electronic data interchange and management information software are all examples of the way in which technology has sought to transform the retail operating environment.

In addition, technology has been a central factor in assisting retailers to expand overseas. While the internationalisation of retailing has been well documented and is not a new phenomenon, the scale and regularity of foreign developments have increased dramatically. Tordjman (1994), for example, notes that in 1992 there were more than 1321 international retail establishments in the EC, which compares with 120 in 1970. France, Germany and the UK are among the most progressive countries, accounting for two-thirds of these operations. The control systems made available through technology have reduced the significance of distance as a mediating factor upon strategic growth. Consequently increasing numbers of retailers have sought to expand into new geographical markets through joint ventures, acquisition and merger, takeover, organic growth and franchising (Dawson, 1995).

The traditional view of the retail sector is one of low pay, low skill and low productivity. While undoubtedly many examples of this scenario still exist today, retailers are faced with a dilemma. The competitive nature of the retail environment has meant that labour costs have to be tightly controlled, although against this consumers require higher levels of product knowledge and better levels of customer service. The issue becomes one of maintaining a balance between cost control and service delivery. Some of the more progressive retailers have invested heavily in education and training programmes. Retailers have increased the numbers of graduates employed and management development programmes have become an integral part of career progression.

The Airport Retail Market

Airport retailing lies within the broader ambit of the travel and tourism industry. The sector as a whole accounts for approximately 200 million jobs worldwide and this number is expected to grow to over 300 million by 2005. Output from the industry for 1995 was estimated to be approximately US$3.4 trillion rising to US$7.2 trillion by 2005 (WTTC, 1995). Estimates of the size of the retail travel market remain difficult to gauge primarily owing to the problems experienced in accruing detailed statistical information on sales revenues from such a wide and varied set of distribution channels. The World of Travel Shopping (WTS, 1995) estimates that revenue from international travel shopping was in the region of US$60 billion. The largest proportion of this income comprised duty-paid shops[1] (71.6%), land shops[2] (11.5%) and airport shops[3] (10.9%). The remainder comprised sea shops[4] (3.5%) and sky shops[5] (2.5%).

Duty and Tax Free

ETRF (1996) estimates that the 1995 worldwide duty/tax-free sales through all distribution channels was approximately US$20.5 billion (including sales through military shops and diplomatic channels). Airports, airlines and ferries accounted for US$13.1 billion. Europe remains the largest single market with duty/tax-free sales accounting for US$10.3 billion (50.4% of the worldwide share) (Table 2.1).

Table 2.1 *Duty/tax-free markets disaggregated by sales channel, 1995 (US$ millions).*

	Airports	%	Airlines	%	Ferries	%	Other	%	Total
Europe	4035.3	39.0	1081.3	10.5	2747.1	26.6	2477	24.0	10340.5
Americas	1378.1	37.3	175.1	4.7	15	0.4	2129.8	57.6	3698.1
Africa	107.4	52.1	34.8	16.9	3.7	1.8	60.3	29.2	206.2
Asia and Oceania	3076.2	49.2	475.2	7.6	–	–	2703.8	43.2	6255.2
Total	8596.9	41.9	1766.4	8.6	2765.7	13.0	7370.9	36.0	20500.0

Source: Best 'n' Most (1996).

Table 2.2 illustrates the world's top ten duty-free markets. In 1990 the USA was the country with the largest duty/tax-free sales accounting for just under 12% of market share. Since then, the UK has overtaken the USA. Airport operators in the UK have aggressively expanded their retailing activities, which has helped improve their total level of sales and establish their ranking. The high geographical concentration of

---

[1] Includes travellers shopping in landside airport shops: department stores, hotels, souvenir shops etc.
[2] Includes downtown shops claiming tax-free status without affiliation to airport duty-free shop operations.
[3] Includes duty-free shops at airport shops as well as affiliated off-airport shops.
[4] Includes sales on board ferries and cruise ships excluding sea port shops.
[5] Includes in-flight sales on board aircraft both scheduled and charter.

duty-free sales is also highlighted in Table 2.2 with 49% of total sales being through ten countries. If the top twenty five countries are taken together then this level of concentration increases to approximately 80% (Best 'n' Most, 1996).

Table 2.2   *Sales in the world's top 10 duty/tax-free markets, 1995 (US$ millions).*

Rank	World	Sales by value	Market share
1	UK	1827	8.9
2	USA	1447	7.1
3	South Korea	1052	5.1
4	Germany	1019	5
5	Finland	936	4.6
6	Hong Kong	889	4.3
7	Japan	798	3.9
8	France	718	3.5
9	Denmark	706	3.4
10	Netherlands	668	3.3
11	Others	10440	50.9
Total		20500	100

Source: Best 'n' Most (1996).

Airport retailing is often divided into two segments: airside and landside. Airside retailing covers those outlets that operate after passport or security control and have traditionally represented the primary retail revenue source for airport operators. Landside facilities relate to all areas that may be accessed by members of the public and offer a wider range of duty-paid goods and services to the consumer. While retail outlets on the airside are mainly duty/tax-free and therefore offer a price advantage to the consumer, they are not exclusively so. For example, while duty-free goods are sold airside in Germany, law prohibits the sale of tax-free products. In the UK, books/magazines are product groups that are exempt from tax and have the same selling price throughout the airport.

Identifying the total retail sales within an airport remains difficult owing to the lack of accurate information on the sale of duty/tax-paid goods. While landside shopping is increasing in importance it still remains a relatively small proportion of the total level of retail sales within an airport. Duty/tax-free sales remain the most significant source of retail revenue accounting for approximately 68% of all retail sales within an airport, compared with 19% for news and books, and 14% for duty and tax paid (Humphries, 1996).

Retail sales through airports are estimated to be approximately US$8.6 billion in 1995 with the largest proportion of this spending being in Western Europe. European airports account for almost half of this with US$4 billion or 47% of global duty/tax-free sales. While the European market continues to dominate total airport retail sales and duty/tax-free purchases, both the Americas and the Asian Pacific market have shown higher levels of sales growth (Table 2.3). Best 'n' Most (1996) estimates that total trade will increase by 9.5% and 8.4% respectively in these regions. Africa, although showing significant year-on-year growth between 1990 and 1993, is estimated to account for only 1% of total duty-free sales.

Table 2.3 *World airport duty-free sales by region, 1990–1995 (base 100).*

	1990	1991	1992	1993	1994	1995
Europe	100	95.07	105.45	112.69	123.73	165.37
Asia Pacific	100	105.80	113.36	124.49	145.36	159.3
North and South America	100	89.96	109.4	115.2	129.03	173.72
Africa	100	147	171.6	177.35	111.32	149.34

Sources: various.

Despite having 46% of the world passenger traffic in 1994 and having the second highest level of duty/tax-free shopping, US airports remain a relatively under-utilised market channel. For example, in 1994 airport duty-free sales accounted for only 27% of total retail sales, less than half the global average (Humphries, 1996). One reason for this is the high proportion of domestic traffic compared with Europe (88.9% in the USA compared with 37.5% in Europe). Also, as will be illustrated in later chapters, one of the factors that influences an individual's propensity to buy duty/tax-free items is the potential savings that can be made relative to purchasing in the home country. Some countries such as India impose up to 290% import duties on certain wines and spirits. In contrast the relative cost of similar products in the USA remains low and partly accounts for the limited purchase of alcohol by US passengers travelling abroad.

While Europe remains the largest region for duty-free sales, it accounts for only four of the top ten duty-free outlets (Table 2.4). A number of reasons have been put forward to account for this, including the growing popularity of other regions as holiday destinations, the rise in international business traffic, a desire for western branded products throughout Asia, an improvement in the range and quality of duty-free retailing, prohibitive liquor and tobacco taxes, and the custom of bringing back presents for friends and relatives (WTS, 1995; Humphries, 1996).

Table 2.4   *The world's top 10 duty-free outlets, 1995.*

Rank	Outlet	Sales (US$ m)	Sales/pas. (US$m)
1	London Heathrow	524.1	22.5
2	Honolulu	419.4	109.83
3	Kai Tak Hong Kong	400	25
4	Changi Singapore	358.8	15.47
5	Schiphol	326.6	26.13
6	Manila	302.6	32.92
7	Frankfurt	289.4	19.99
8	Paris CDG	Conf.	Conf.
9	Tokyo Narita	280	25.45
10	Guaralhos Sao Paulo Brazil	Conf.	Conf.

Source: Duty Free Database and Directory (1996/97).

Duty-Free Abolition

One of the most significant factors due to influence duty/tax-free sales is its proposed abolition on intra-EU travel. As illustrated in Chapter 1, the intended date for this to occur is June 1999. The impact will not be limited to airports and will extend to the three primary sales channels.

Table 2.5   *Estimated loss from the abolition of intra-EU duty/tax-free sales.*

	Current sales			Intra-EU loss		
	US$ bn	ECU bn	%	US$ bn	ECU bn	%
Airports	3.4	2.6	49	1.8	1.4	52
Ferries	2.5	2.0	37	2.4	1.9	97
Airlines	1.0	0.8	14	0.5	0.4	56
Total	6.9	5.4	100	4.7	3.7	69

Source: ETRF (1996).

While all three main distribution channels for duty free are likely to be adversely affected, the ferries stand to suffer the largest percentage revenue loss from the abolition (Table 2.5). Owing to the limited number of European-based ferries travelling to non-EU destinations, the percentage decline will be disproportionately greater. Because of the greater reliance of some countries upon ferry services, the impact will have a geographical dimension (Table 2.6).

Table 2.6 *The impact of duty/tax-free abolition by country, 1995.*

Country	Total Sales		Intra EU		
	US$ m	*ECU m*	*US$ m*	*ECU m*	*% Loss*
Austria	70.3	55.1	44.2	34.7	63
Belgium	180.0	141.1	130.0	101.9	72
Denmark	718.7	563.5	570.3	447.1	79
Finland	901.0	706.4	813.6	637.8	90
France	634.9	497.8	384.9	301.8	61
Germany	776.2	608.5	477.9	374.7	62
Greece	180.2	141.3	115.2	90.3	64
Ireland	146.9	115.2	116.7	91.5	79
Italy	279.8	219.3	152.7	119.7	55
Luxembourg	15.5	12.2	14.2	11.1	91
Netherlands	421.0	330.3	198.8	155.8	47
Portugal	63.8	50.0	40.6	31.8	64
Spain	346.5	271.6	204.6	160.4	59
Sweden	461.7	361.9	396.8	311.1	86
UK	1678.4	1315.8	1102.5	864.4	66

Source: ETRF (1996).

For example, in Finland and Sweden 89% and 67% of all duty free is sold on board ferries. The consequent loss of duty-free revenue is expected to be in the region of 90% and 86% respectively (ETRF, 1996). A second factor affecting revenue loss will be the number of non-intra-EU flights that an airport has. Airports with flights to other international destinations will continue to sell duty free. The Netherlands, for example, is used by passengers as a transit hub to the Middle and Far East. Despite 86% of its duty/tax free being sold through airports it will expect a 47% loss of revenue. This may be compared with both Ireland and Luxembourg. Like The Netherlands, a high proportion of duty/tax free is sold through the airports (81% and 95% respectively). Unlike The Netherlands, however, both countries have a much higher incidence of

Table 2.7 *The main players in the duty-free market, 1995/1996.*

Name of operator	Turnover US$m, 1995	Market share (%)	Main markets	Main activities and interests
Duty Free Shoppers (DFS)	2800	13.7	US Pacific Region	The world's largest duty-free group. Originally founded by Charles Feeney. The company have over 180 outlets trading in 11 countries. 61.25% share bought by Louis Vuitton Moet Hennessy (LVMH) 1996
Nuance/Allders	1159*	5.9	Europe, UK, Australia	Established in 1992 through a merger between Crossair and Swissair duty free. By end of 1995 had 13 airport shops. Acquired Allders in July 1996. The company also owns City International and Downtown Duty Free companies in Australia
Gebr. Heinemann	800	3.9	Germany, Spain, Portugal, USA	Europe's largest duty-free retailer. The company operates eleven airport duty-free shops in Germany. It also has a 50/50 partnership at Paris Orly South for fragrance and cosmetics. In addition, Heinemann own a duty-free and a diplomatic supply company, border shops and supermarkets
Weitnauer	744	3.6	Europe, Africa, South America	The company operates in over 20 countries through a network of subsidiary companies and joint ventures. It runs 80 duty-free outlets in airports, border crossings, seaports and cruise liners
Duty Free International †	515	2.5	USA	One of the leading airport duty-free operators in the USA. DFI has 85 duty-free outlets in US and the Caribbean. It has border shops in both the north and south of the USA and also a separate diplomatic division

Stena Line	493	2.4	Scandinavia	A Swedish-based ferry service that operates 34 ferries across 16 routes sailing from Scandinavia, the Baltic and the UK
Alpha Retail Trading	481	2.4	UK	ART specialises in in-flight catering and retailing. It operates at 22 UK airports and runs 31 duty-free shops and 52 tax-paid stores (principally bookstores, chemists and gifts)
Silja Line	402	1.9	Baltic	The company has 19 passenger ships in the Baltic, operating primarily between Sweden and Finland. It has approximately 19,500 berths and carried 5 million passengers in 1995. Silja also owns Sally UK, the cross-channel ferry operator
Aer Rianta	355	1.7	Ireland, Russia, Middle and Far East	The airport authority in Ireland who is also responsible for duty/tax-free retailing in Dublin, Shannon and Cork airports. The company's international arm has joint ventures and management contracts in Russia, Bahrain, Karachi and Beijing
Duty Free Philippines	336	1.6	Philippines	A Government-owned operation based at Manila airport with ten other stores nationwide. The company principally focuses upon selling duty/tax-free goods to returning Filipino workers.
SAS Trading	300	1.3	Scandinavia	Part of the Scandinavian Airline Systems (SAS) group. The company has 40 shops in 20 airports in 6 countries and is responsible for all in-flight sales on SAS airlines

* indicates the combined turnover of both companies prior to takeover.
† Prior to the takeover by BAA, July 1997.

intra-EU flights and will consequently suffer significantly larger losses in EU revenue (79% and 91% respectively).

Duty/tax-free retailing represents a complex industry where the boundaries between retailer, wholesaler, airport operator and manufacturer have become increasingly blurred. Table 2.7 highlights the main organisations involved in duty/tax-free retailing. Companies such as Heinemann are retailers, distributors and agents; Aer Rianta is both a retailer and a manager of airports; LVMH is a retailer and a manufacturer as well as a supplier to the competition. The level of internationalisation remains high and the extent to which companies are willing to look outside their national boundaries has been intensified with the proposed abolition of duty free on intra-EU flights.

Product Demand

Within the duty-free industry, liquor, fragrances/cosmetics and tobacco dominate the worldwide sales market. Airports represent the single most important distribution channel for the sale of these items. In 1995, 36.8% of wine and liquor, 42.4% of fragrances and cosmetics, and 46.9% of tobacco sales were sold through airports (Table 2.8)

Sales in all three principal product categories have continued to demonstrate year-on-year growth. For example, between 1994 and 1995 sales of cosmetics were stimulated through new product launches and grew by 16%. Liquor sales increased by 11.1%, boosted in part by product line extensions and exclusive duty-free products. Tobacco sales also increased by 8.7% (although their share of the total world market dropped from 13.1% to 12.4%) while gifts and other goods also grew by 16.6% (Lloyd-Jones, 1996). As Table 2.9 illustrates, Europe has the greatest demand for each product group, for example in the fragrance and cosmetics category, sales increased by 20% in 1995.

Product demand is changing however. The long-term viability of tobacco as a major revenue and profitability contributor remains open to question. Some such as Simon (1993) and Campbell (1994) note sustained growth in the sale of tobacco while others such as Gibson (1992) maintain that in future, liquor and tobacco cannot be relied upon to contribute to profitability at the same rate as in the past. Consumers are moving away from products such as cigarettes and spirits towards wines and beers. While consumer demand may be in a period of transition, this change has not been influential enough to fundamentally alter the demand for the most popular product categories (Table 2.10).

Table 2.8 Product demand by distribution channel, 1995 (US$ millions).

Channel	Wines and Spirits	%	Perfumes and Cosmetics	%	Tobacco Goods	%	Misc. Goods	%	Total	%
Airport shops	1997.8	36.8	2173.3	42.4	1200.8	46.9	3225	43.7	8596.9	41.9
Airlines	321.9	5.9	620.6	12.1	253.3	9.9	570.7	7.7	1766.4	8.6
Ferries	901.9	16.6	325.5	6.4	510	19.9	1028.3	13.9	2765.7	13.5
Other shops and sales	2212	40.7	2005.8	39.1	598.9	23.4	2554.2	34.6	7370.9	36
Total World	5433.6	100	5125.2	100	2563	100	7378.2	100	20500	100

Source: Best 'n' Most (1996).

Table 2.9 *Product demand disaggregated by market, 1995.*

Country	Wines and Spirits	%	Perfumes and Cosmetics	%	Tobacco Goods	%	Misc. Goods	%	Total	%
Europe	2571.6	47.3	2666.9	52	1725.6	67.3	3376.5	45.8	10340.5	50.4
America	1208.8	22.2	1059.5	20.7	286.9	11.2	1142.9	15.5	3698.1	18
Africa	56.2	1	50.7	1	36.1	1.4	63.3	0.9	206.2	1
Asia & Oceania	1597	29.4	1348.1	26.3	514.5	20.1	2795.6	37.9	6255.2	30.5
Total World	5433.6	100	5125.1	100	2563	100	7378.2	100	20500	100

Source: Best 'n' Most (1996).

Table 2.10   *Rank order of top 10 product categories bought through tax and duty free.*

Rank	Product category	1995 Sales (US$ millions)	Market share (%)
1	Women's fragrances	2250	11
2	Cigarettes	2235	10.9
3	Women's cosmetics	1767	8.6
4	Scotch whisky	1663	8.1
5	Cognac	1235	6
6	Men's fragrances & toiletries	1109	5.4
7	Accessories	1050	5.1
8	Confectionery	1049	5.1
9	Leather goods	902	4.4
10	Watches	703	3.4
Total top 10		13962	68.1

Source: Best 'n' Most (1996).

Typologies of Airport Retailing

As one may expect from a global industry, there remains no single approach to retailing within an airport. Differences exist in the priority accorded to commercial activities as well as the competency of the operators to develop non-aeronautical revenues. The previous chapter discussed the operation of airports and the function of retailing within this context. While Doganis's (1992) model of a traditional versus commercial airport remains a useful starting point, this basic distinction may be augmented through the development of four basic typologies of airport retailing. As is the case in any such classification, exceptions to the rule can be listed, as well as hybrid forms identified. A relative rather than an absolute interpretation is therefore necessary.

Type 1:   Concessionaire-Based Retailing

Concessionaire-based retailing represents the dominant form of retail operation within airports and occurs when the airport operator plays a strategic rather than an operational role in the retail activities. Responsibility for the sale of goods lies with third parties who have been contracted in by the authorities and whose terms and conditions are negotiated periodically. The airport operators occupy the role of landlords and are responsible for the physical facilities of the building and the provision of the contracted service requirements (e.g. heat, light). Some airport authorities have taken a more proactive role and attempted to develop an environment within the airport that is conducive to retail activity (see Chapter 4). While the airport operator has no involvement in the purchase or sale of product, there may be a commitment to market the airport as a whole. A Type 1 operation is illustrated below (Figure 2.1). Examples of this approach to airport trading include Schiphol, Frankfurt, Copenhagen and Paris Charles de Gaulle.

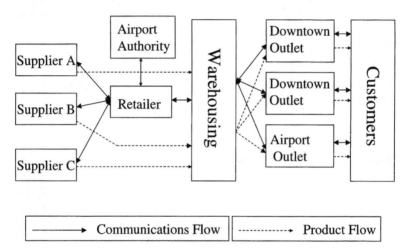

Figure 2.1 *Consessionaire-based retailing.*

For the airport operator the objective remains clear, to maximise the amount of revenue gained from the concessionaires while at the same time maintaining a level of customer service consistent with the airport's overall business objectives. The majority of airports have some form of concession arrangement and the generation of revenue by this method represents one of the most widely adopted forms of commercial contract. Airport operators will seek to utilise the available floorspace to its maximum potential by offering concession arrangements to a range of service providers including retailers, car hire firms, caterers and even hotels.

For the retailer the advantage of operating in an airport lies in the large passenger flows to which the business is exposed and the high potential turnover this generates. For example, the largest outlet for Rolex watches in the UK is in Heathrow Terminal 4. Furthermore, Shaw

(1993) notes that Bally, the international shoe retailer, had a revenue return of between £2300 and £2600 per square foot in Heathrow as compared with £250 per square foot in the high street. The rental paid to BAA was £270 per square foot or £1.35 million for a 5000 square foot unit.

Doganis (1992) maintained that there were a number of factors that could influence the total amount of revenue generated by a concessionaire. The first is the total passenger traffic within the airport. Many airports have seen the volume of passenger traffic grow significantly over the past two decades and have consequently benefited from the increased customer spend that this has brought. However the extent to which an airport operator is able to influence this flow remains limited. Some attempt can be made to attract new carriers to use the airport, but little control can be exercised over the volume or composition of the passenger traffic. For the airport operator, therefore, the main objective will be to ensure that it establishes a reputation for professionalism and the ability to cope with the fluctuations in passenger demand that are characteristic of many airports.

Concessionaire tendering

The process of tendering for a concessionaire will depend upon the size and type of airport, the amount of retail space available, the experience of the airport operator in managing tenders and the market power of the individual retailer's tendering. It is therefore difficult to provide a definitive description of the tender process. What will be outlined here is an approach towards best practice as identified by the industry, while at the same time acknowledging that many deviations and idiosyncrasies from this description exist.

The requirements of each airport will vary and operators have wide discretion as to the terms and conditions that they can impose upon bidders. Placing concessions out to tender however remains the most widely adopted method of utilising commercial space within airports. Tenders are advertised in a number of ways, including the trade press, national newspapers or even word of mouth. A retailer's decision on whether to tender will be based upon a number of financial and non-financial criteria. Perhaps the most important will be whether the concessionaire will be able to run the stores profitably over the period of the contract. If the financial return demanded from the airport operator is too high or the conditions imposed upon trading remain too stringent then this may not be possible and a retailer may decide not to submit a bid.

Deciding whether to tender for a concession is not always based upon the objective of maximising financial returns. A strategy may be determined by a number of other evaluative criteria; for example, the extent to which the retailer wishes to be represented in a particular country or the need to portray a global image. Alternatively it may be considered necessary to gain a foothold in a country or region that displays significant growth potential. The increase in Pacific–Asian travel over the past decade highlights the extent to which emerging markets can represent highly profitable revenue streams for retailers with proactive globalisation strategies.

The tendering process itself varies. Some tenders are open competition, which means that any individual or organisation is able to apply for the concession. A potential problem of this strategy in the past has been the large number of bidders who have put in unrealistic tender offers and have experienced difficulties in managing and running the concession at a profit. Some operators have sought to avoid this issue by requiring all potential bidders to meet certain criteria before being allowed to place a formal tender. This *pre-qualification* stage requires the bidder to provide information about its organisation, including its previous track record, its financial base, the number of concessions already held and the expertise of the existing management. In addition the ability to demonstrate an understanding of the demands of working within an airport environment is also desirable. The airport authority then screens these applications and, if successful at this stage, the organisation is then provided with all relevant documentation and allowed to place a formal tender for the concession. The *tender* stage requires the potential concessionaire to submit a detailed marketing and financial plan.

The marketing plan covers the main operational procedures that the concessionaire would have in place if its bid were successful. In particular it would detail how the shop was to be designed and merchandised, what products would be sold, the price levels of the merchandise, what promotions would be initiated, the personnel requirements and the types of training that would be conducted. Table 2.11 highlights the broad headings contained within a concession document and the areas to which a bidder would be expected to respond.

The financial plan details the fees offered to the airport by the bidder. The form that this takes depends upon what is required by the operator. At its simplest it may comprise a single percentage rate for all product groups that the retailer is intending to sell, alternatively the operator may require separate percentage returns across different product groups. In one recent example the airport required all those tendering to provide five year sales projections and fee offers for 52 separate product groups. A retailer may be obliged to provide more detailed financial projections based upon differing levels of sales. As turnover rises so the fees payable to the airport operator may also be expected to increase.

Table 2.11 *Main headings and sub-headings for a tender document for a retail concessionaire.*

Section A	*General Information:* Basic location of airport and hinterland access, basic information on airport including numbers employed, passenger profile, peak hours, outlook for passenger traffic
Section B	*Present Airside Shopping Facilities:* What shops currently operate in the airport, where they are located, how access is gained to the retail area, what merchandise they are selling, what legislative requirements exist airside and landside
Section C	*Alteration of Facilities:* What alterations are allowed to the basic store design, who pays for alterations, who retains ownership of fixtures and fittings on termination of contract, if layout of terminal changes during concession period what are the obligations on part of the airport
Section D	*Minimum Service Required:* Opening hours, number of days open p.a., staffing levels, basic stock and duty-free ranges, quality of furnishings, provision of basic statistical agreement
Section E	*Contract Conditions:* Outline of Heads of Agreement, including Length of Service, Period of Agreement, Payment Details, Hours of Operation, Areas of Airport Access, Insurance and Liability Requirements, Security Arrangements etc.
Section F	*Tender Details:* Period of agreement, guaranteed minimum payment, projection of gross revenues, concessionaire fee paid by each merchandise category, submission of designs and layout, projection of gross revenues, proposed selling price, daily, weekly and seasonal staffing levels, financial information required, wage levels. How to submit the tender
Section G	*Drawings:* Surrounding region, airport itself, layout of terminal, layout of store and shopping areas
Appendix 1	*Statistics (Passengers)*
Appendix 2	*Schedule of Existing Fixtures and Fittings*
Appendix 3	*Correspondence with Customs (Bonded Store)*
Appendix 4	*Customs and Excise leaflets – Operating a Duty Free Shop and Customs Warehousing*
Appendix 5	*Details of Turnover of Existing Shops*

While such an approach has merit, a prerequisite for the airport operator must be to provide clear financial guidelines for tenders to be compared against. One issue that has occurred under this latter method is that less experienced airport operators have developed overly complex methods of determining concession fee revenue, making it difficult to evaluate comparable bids from retailers.

The decision on which tender to accept is made by the airport operator. The most equitable means of making this choice is to open all sealed bids at the same time (ideally with representatives from all tendering parties present). The projections provided under the financial plan represent the key factor in deciding whether a bid is successful and usually takes precedent over other factors. In order to ensure that the marketing plan is considered, some airports require the financial and marketing bids to be submitted separately. The marketing bids are opened first and evaluated. Only if these are considered to be competent and workable are the financial bids opened.

The time period between the announcement of a tender and conveying the final decision to the applicant will depend upon the number of tenders placed and the size and scale of the unit on offer. However as a broad guideline of best practice, the bidder should be given a period of between one and two months to submit the pre-qualification document. The airport operator should respond to this submission within six to eight weeks and, if successful, allow the bidder access to the tender documents. (Tender documents are sometimes sold by the operator as a way of gaining additional revenue.) Drafting and submitting the tender document should then take another two to three months. If the retailers are present at the opening of the sealed bids then they will know immediately if their tender has been successful. If the airport operator decides that no bidder should be present then all interested parties should know within two months.

Correct choice of contract

Developing a contract that provides both parties with mutually beneficial terms and conditions remains a central objective for the operator. If terms and conditions remain too stringent and limit the retailer's ability to operate, then there is likely to be an under-performance in trade. Conversely, the airport operator has responsibility for the retail offer across the entire airport and must ensure that any concessionaire has the professionalism and expertise to deliver to agreed standards. The role of the operator in this context is twofold. First, to provide a degree of co-ordination and continuity between the different types of retail offer. Secondly, to draw a balance between the creation of a stable operating environment while not becoming tied to long-term inflexible arrangements. The correct choice of contract is therefore essential for the successful running of a retail concession.

The duration of each contract and the financial conditions attached to each will vary by airport and retailer. It is unusual to have specified rentals for different parts of the terminal and, unlike the high street, the retailer in an airport has no automatic right to a rent review. The length of the contract is dependent upon a number of criteria. One of the most important is the level of investment required from the retailer. If little investment is required then a contract is often short term. Papagiorcopulo (1994), for example, notes that concessions in Malta's International airport are offered on a one year renewable basis. The rationale for this is to ensure that concessionaires remain competitive. A similar situation existed in Schiphol until 1991. Although concessionaires were not required to tender every year, trading terms and conditions had to be negotiated on an annual basis. For the retailer such requirements have little attraction. If any significant level of investment were required from the retailer then a contract of 5 years would be considered the minimum. Doganis (1992) maintains that contracts should not exceed five years without the opportunity for other retailers to tender for the concession. Seven year contracts however have become more widespread for larger retailing groups and examples of contracts in excess of ten years are not unknown.

At the end of the contract, a provision may have been built in to extend the agreement for a further pre-defined period. The retailer may then continue to trade under the same terms as previously provided, but more commonly, however, the contract is extended after new terms and conditions have been negotiated. If the two parties are unable to reach agreement then the contract may be put out to re-tender. Amsterdam's Schiphol airport, for example, has a policy of renewing existing concessionaire contracts provided they have met the required pre-set criteria laid down by the operator. This method has a number of advantages in that it encourages the creation of good working relationships between the two parties and allows the concessionaire to develop a degree of competency in the field of airport retailing. Such an approach however avoids the potential benefits that accrue from a competitive bidding process. The operator may accept lower fee income than could be achieved under a tendering process and lose the opportunity to provide new, more innovative types of retail services.

The extent to which the airport operator exerts control over the product market varies. Schiphol, for example, has a strict non-competition policy between concessionaires. The merchandise that can be sold in each outlet is agreed and regulated by the airport authority. Where two competing retailers have each won a tender to sell the same products in different parts of the terminal, e.g. photography, then both must follow a policy of price parity. In contrast, other airport operators have adopted an alternative approach. Doganis (1992) notes that some airport authorities actively encouraged competition between retailers on the basis of providing greater choice to the consumer and reducing the criticism of them exploiting a monopoly position. For example, BAA views each of the airport terminals in which it operates as separate markets and promotes a strategy of direct competition between all concessionaires. BAA maintains that this both improves standards as well as keeping prices under control. At Heathrow Terminal Four this strategy has been taken a stage further with direct price competition existing between duty/tax-free operations (Gibson, 1992).

Concessions pricing and fee structure

The negotiating price between concessionaire and operator remains a delicate process. Airport operators have the responsibility to seek a maximum return on behalf of their stakeholders; retailers on the other hand are required to contractually commit themselves to paying an agreed level of revenue on the basis of speculated traffic flows. Their objective is to ensure that they have as much relevant information as possible before the tender is submitted. Few airports accept responsibility for imprecise passenger estimates and the contract between retailer and operator remains binding regardless of the actual numbers travelling. When deciding upon the amount to tender the retailer will consider a variety of factors. These include the number of passengers travelling, the passenger mix, the destinations served by the airport, the number of persons in the terminal at the same time, the location of the shop, the length of the contract, the size of the unit, the controls placed upon product assortment, and the legislative and fiscal controls placed upon retailers and travellers.

Some airports such as Malta International initially began by leasing/renting out space to retailers for a fixed fee. In addition to the difficulty of maintaining control over this form of operation, an added disadvantage of this agreement is the potential loss of retail revenue. Regardless of the success of individual units within the terminal the revenue derived from this form of tenancy contract remains fixed. In contrast a more widely adopted method of payment for a concessionaire is to link fees directly to the financial turnover of the outlet. As sales increase the amount payable to the operator also increases (Table 2.12). In this way both parties benefit from the retail operation.

Table 2.12   *Singapore Changi terminal one perfume and cosmetics tender result (US$ millions).*

Bidder	1994/95	1995/96	1996/97	Total
SADE-DFS	13.0	14.8	16.3	44.1
Scotts Weitnaur	11.4	12.0	12.6	36.0
Allders International	8.9	10.8	12.5	32.2
BBH	5.8	5.9	6.0	17.7

Source: Duty Free Data Base and Directory (1994; p. 9).

All bidders offered percentage sales guarantees of 30% of sales if annual minimum guarantee levels are exceeded.

Using turnover-based rents as a means of concessionaire payment provides the operator with a number of benefits. First, it has the potential to provide a high net return (while figures as low as 5% are levied on some outlets, returns of between 45% and 50% of turnover are not uncommon on the more profitable airside concessions). Furthermore, as sales increase the percentage fee payable to the operator may increase disproportionately to the level of turnover. The rationale for this is based on the retailer having covered fixed and operating costs and therefore being able to provide the airport with a higher percentage contribution. In addition to the sales-based turnover fees, the operator will ensure a basic return from each concession by specifying a minimum fixed rental fee. Thus even if the retailer performs well below its forecasts, the airport operator has a guaranteed income.

Concessionaires also represent a relatively low financial risk to the operator. While some airport authorities traditionally provided all shop fittings and helped in the construction of the retail units, today such a strategy is much less common. Responsibility for merchandising, store layout and operations in most instances lies with the retailer (Gibson, 1992). Concessions also remain relatively easy to administer. For example, competitive bids against pre-set criteria allow easy comparison, while the collection of rental income remains a straightforward calculation. The operators main financial risk lies in the opportunity cost of running a poorly performing outlet and any consequent knock-on effect this may have upon sales in other outlets.

There are however a number of disadvantages associated with running a traditional concessionaire agreement within an airport. Livingstone (1995), for example, notes that the retailer who puts in the highest bid to the airport is not necessarily the retailer who will provide the highest overall revenues to the operator. Even where the operator specifies a pre-qualification stage, retailers may provide unrealistic sales projections that cannot, in practice, be achieved.

By using concessionaires the operator relinquishes direct control over the product categories stocked within each store. While there may be strict guidelines ensuring that no two stores stock similar merchandise (as at Schiphol), the operator is unable to directly influence the actual choice of merchandise. Retailers may therefore seek to maximise margins rather than sales turnover by stocking only the most profitable product categories. Furthermore the concessionaire formula provides little incentive for high levels of sales-based performance. Where competition is encouraged between retailers, price promotion may further undermine the revenues accruing to the operator. Such strategies limit total revenues, the selection of merchandise on offer and may even compromise customer service as retailers attempt to control employment costs.

Type 2: Authority-Managed Retailing

While concession retailing represents the most widely practised form of retailing, other methods of operation also exist within airports. With a Type 2 method, direct responsibility and control are assumed for a proportion of the retailing by the airport managers themselves. Their remit therefore extends beyond that of landlord and administrator to the active participation in the airport's commercial strategy. Primarily this strategy focuses upon retailing although the commercial strategies may extend to include catering and car parking (Keogh, 1994). The operator will primarily focus upon duty/tax-free products although the organisation may also operate landside retail facilities. The buying of product from suppliers becomes the direct responsibility of the airport operator. Product is delivered to the airport warehouse and all merchandising, marketing and stock control remain the remit of the airport operator. Concessions are still used within the airport although they do not represent the primary method of retail development and may be used to supplement the portfolio of goods and services on offer. The product range offered by concessionaires tends to focus upon areas where the airport management has decided not to compete; for example, where a high degree of expertise or product knowledge is required, as in floristry or fresh goods retailing. Figure 2.2 illustrates an authority-managed retail operation in an airport. Aer Rianta, the Irish airport management group based in Dublin, is an example of this type.

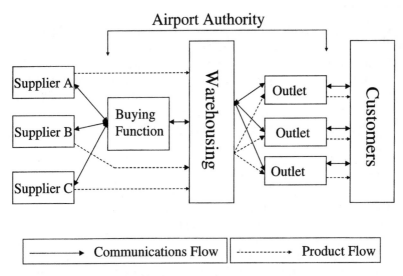

Figure 2.2   *Authority-managed retailing.*

The advantages of this form of operation from the airport authorities perspective are that:

- All revenues accrue directly to the authority. With the increased volume of passenger traffic highlighted in Chapter 1, this has represented a highly lucrative market which airport operators have sought to exploit.
- An authority-managed operation provides direct control over the goods stocked and the way in which they are merchandised in-store. This allows the retailer to concentrate upon achieving the correct product mix rather than focus purely upon sales or margin.
- Consistency between different advertising media can be achieved. Typically an airport has to draw a balance between information provision and marketing communications. The latter is provided in-store, in-terminal and externally using a variety of methods including signage, billboards, TV and mailings. An authority-managed operation

has the opportunity to administer this task centrally and thereby provide a greater degree of overall consistency and control.

- Pricing can be set centrally to ensure revenue maximisation.
- Customer service standards are more likely to remain consistent throughout the terminal building.

There however remain two significant disadvantages of using a Type 2 operation. These are:

- The airport authority is exposed to a greater degree of risk than if a purely concession-based strategy is followed. The operator may not have the necessary skills required for airport retailing, nor an understanding of the merchandise requirements of its customers;
- A significant set of financial outlays, not only for merchandise and sales staff but also for fixtures, fittings, warehousing and specialist personnel (buyers, merchandisers, space planners etc.).

Type 3: Management Contract

Management contracts exist when a third party organisation is contracted to operate elements of the retail operation on behalf of the airport. Unlike concession retailing, this form of relationship requires a greater degree of commitment and partnership between the two parties. If the operator of the airport does not have the necessary retail skills or experience, an existing retail group may take responsibility for retailing within the airport in return for a pre-determined set of fees. Management contracts can operate in a variety of ways, for example on a full cost recovery basis, whereby the operator's costs are met in full and a management fee is provided that is linked to a pre-determined level of sales. Both parties agree a sales and profit budget, and incentive arrangements are built in to encourage the contractor to exceed the financial objectives set.

In some instances management contracts do not exclude the airport operator from having some involvement in the retail operation. For example, prior to the development of World Duty Free, BAA changed its duty-free operation from a concession system to a management contract. BAA began purchasing all stock, retaining all sales revenue and taking all responsibility for the upkeep of the outlets. In return for operating the outlets, the retailer earned a series of graduated management fees linked to specified levels of sales turnover. Merchandise selection and pricing were administered under a joint board of BAA and retail executives. Under this system greater control is exercised over the total retail offer by the airport operator, while at the same time retailers are provided with the necessary incentives to maximise sales rather than focus purely upon margin.

An alternative and less common form of management partnership is the development contract. Here the airport operator decides to lease out its entire retail space to a third party. It is the remit of the third party to develop, build and manage all the concessions and to control the occupancy of retail space within the airport. The BAA operation in Pittsburgh International airport works under this form of relationship. In return for a guaranteed sum paid to the airport managers (40 cents per passenger travelling), the company has responsibility for developing and running retail operations at the Midfield Terminal. Under a management contract, many of the problems experienced under a Type 1 concession arrangement are avoided. While the financial risk to the airport operator is increased, this may be offset against the advantages that accrue under such a relationship. In particular, a closer working relationship exists as both parties share common objectives. Gray (1994) also maintains that a management contract provides a greater degree of security of tenure, the temptation for the retailer to focus upon margin rather than sales turnover being avoided. The incentive to concentrate upon a limited product range is also removed and the airport can focus upon developing a wide and balanced merchandise mix. Figure 2.3 illustrates a form of management contract.

Type 4: Joint Venture Retailing

One of the strategic options discussed in Chapter 1 was the expansion of foreign activities by airport operators. Partly in response to the threat of duty-free abolition on intra-EU flights and the increasing pressure upon European airports to maintain competitive airport charges, there has been an increase in the number of airport operators looking to manage and operate the commercial and aeronautical activities of airports outside their host country. Because retailing represents an activity that offers significant growth opportunities, airport operators have found themselves in direct competition with other market players. Specialist duty-free retailers, airlines, state development corporations and private sector conglomerates are also seeking to increase their retail activities by expanding into airport retailing.

One increasingly common strategy undertaken by airport authorities has been the development of joint venture partnerships and alliances. Western European airport operators, for example, have expanded into

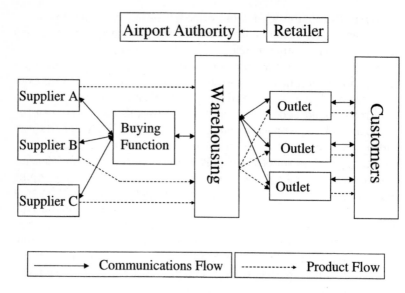

Figure 2.3   *Management contract.*

Eastern Europe, the Middle East and the Far East via contractual agreements with local operators in the host nation. Thus Aer Rianta has been involved in trading with foreign partners since 1988 when it established a joint venture company in the Commonwealth of Independent States (CIS). The objective of the company was to develop duty-free retailing in a number of Russian airports.

The form that these joint operations take and the negotiated trading arrangements under which they function will obviously vary. Typically the airport operator will have its own Overseas Trading Subsidiary (OTS) whose sole responsibility is to manage such alliances. This strategic business unit may then have a further series of established subsidiaries that are responsible for the control of the joint venture within a specified country or region (Figure 2.4). The OTS may use its combined buying power to negotiate with suppliers on behalf of all the subsidiaries and in some instances may attempt to consolidate delivery through centralised warehousing.

Joint ventures are not confined solely to the airport operator. Retailers in their attempts to expand overseas have also sought to develop alliances and partnerships. Allders, the duty/tax-free retailer, for example, had a 50% stake in a joint venture company, Net Magaza Isletmeciligi Ve Ticaret AS in Turkey, to run outlets in five airports.

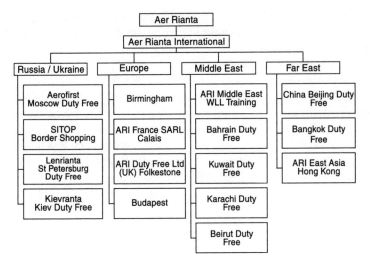

Figure 2.4   *Partial group structure of Aer Rianta, 1995.*

The rationale for developing strategic alliances will vary. Some may have a political dimension; in countries such as Indonesia, strategic alliances with a company from the host nation is a prerequisite to trade. It may also represent a risk-reducing exercise between two organisations as it reduces the competition and avoids either party having to compete against the other. Joint venture agreements provide a number of advantages for both parties involved. For the foreign organisations it allows them to draw upon local knowledge and experience and provides access to an existing infrastructure and distribution network. It can also assist in career development by broadening the staff skill base through foreign work experience. For the local organisation such alliances often provide exposure to a much larger, more experienced organisation which is able to provide specialist expertise and training, often with advanced technology. It may also provide access to previously unobtainable goods, raw materials and foodstuffs, and expand the potential to earn hard currency revenue (Keogh, 1994).

There remains a series of disadvantages that make joint ventures unattractive to many organisations. In particular, conflict can arise over a number of issues including levels of investment, speed of growth, management styles and cultural differences. In addition, a potential loss of freedom can develop from being tied to an individual organisation. The partnership may become strained if one of the members fails to deliver to the pre-required standards and it is not possible to appoint an alternative partner.

Airport retailing must be placed within the context of wider industrial and economic change. It is required to cope with pressures from both the air transport as well as the retail industry. Such demands influence the structure and form of retailing within airports. The liberalisation of air transport, increases in passenger numbers, the threatened abolition of duty free, retailer concentration and expanding internationalisation have created a set of imperatives that demand a proactive response on the part of the industry. In the following chapters the aim will be to take an operational perspective and examine the methods employed by both airport operators and retailers to ensure their sustained existence in what is an increasingly dynamic area of retail change.

3 Retail Location, Planning and Design

Introduction

The planning and design of a new airport or terminal remains a significant undertaking and despite the movement toward a more liberalised air transport market, such developments continue to require the sanction of a number of different bodies. This may range from local or national town planning authorities to central Government authorisation.

In the planning process, airport authorities are obliged to balance a wide variety of influential and mitigating factors as well as account for the interests of different pressure and consumer groups. Under EU legislation there will be a requirement for an environmental impact study to be made to assess such issues as noise, pollution, congestion as well as the impact that the proposed changes will place upon the existing transportation infrastructure. Given the scale of most airport developments such a requirement will apply in the majority of situations.

The impact that a major airport development will have on the local communities must be assessed and their support secured where possible. Proposed schemes are subject to intense scrutiny and are increasingly likely to encounter both serious opposition and actual disruption. Environmental groups in particular have sought to oppose new building because of the large land areas required for development. The new airport at Munich, for example, was initially prevented from opening because of the opposition of local environmental campaigners.

A consideration of growing importance to many airports today is the rapidly increasing cost of expansion and the ability to both cover the initial capital investment as well as future servicing. As illustrated in Chapter 1, Governments not only in Europe but also around the world have actively attempted to increase private sector participation in airport management. The Governments of Australia, Argentina and South Africa, for example, are no longer prepared to fund airport capital development. This decision on the part of the state has demanded that serious consideration be given to the revenue-generating potential of any airport from the initial concept stage.

The level of aeronautical revenue to be generated by a proposed airport is fairly predictable (traffic numbers × airport passenger fees + aircraft × landing charges) and will often be constrained either by the charges levied by competing airports or through statutory regulation. Increasingly it is the non-aeronautical revenue-earning potential of a terminal development that will be of crucial importance. Gray (1997) makes the point that 'In many cases (and increasingly), it is this non-aeronautical revenue which is the very justification for a new terminal' (p. 2).

While the planning issues that surround the construction of airports themselves remain outside the remit of this book, the planning, location and design issues relating to the development of retail facilities within an airport, and in particular within an airport terminal, remain the central focus of this chapter.

The allocation of space for commercial and more particularly retail development within an airport cannot be divorced from the wider considerations of general space planning; for example, any new proposal will be constrained by the existing road networks and aircraft access routes. The *Airport Development Reference Manual*, published by the International Air Transport Association (IATA), sets out broad parameters for land use, airfield configuration and environmental impact issues in the context of airport master planning. It details standards and space planning formulas for passenger and cargo terminals, airside apron areas and the general access to and from the airport. Within a new terminal, the location of specific, non-commercial functions has priority. Airports work to internationally recognised standards which set detailed specifications on the amounts of space to be devoted to essential services such as passenger circulation, check in, baggage handling facilities, security requirements, Government controls, facilities for disabled passengers and sign posting.

Given these requirements it is all the more essential that the commercial or retail concession planner 'be involved at the earliest possible opportunity – before the footprint of the building is fixed and certainly before main flows, service cores and service areas are defined' (Gray, 1997, p. 3). The basic footprint of the building is agreed along with locations for the key aviation-related facilities, the services essential to the building itself and where in principle any commercial developments might be located. Once completed, it is the responsibility of the airport authorities to decide upon the amount of space to be devoted to commercial undertakings, the speed with which such developments should take place and where within the terminal building these activities should be located.

Commercial Space Allocation

Despite the expansion of the retail industry during the 1970s, traditional high street operations were not transferable to airport locations. This was primarily due to the limited space available and the considerable cost of rentals. Equally, the limited number of branded niche retailers in the high street meant that there was little demand from passengers to have equivalent products at the airport.

Despite being unsophisticated by contemporary retail standards, duty-free shops remained a novelty for the majority of air passengers. This was primarily because air travel was an infrequent experience for most people owing to the high ticket costs involved. As recently as the early 1980s, air passengers were still focused on the traditional airport duty-free shops and the internationally branded products that they stocked. The move away from the general store concept selling a wide range of goods and services was highlighted in Chapter 2. The growth of specialised retail chains that became a feature of the 1980s was a more suitable format for transference into an airport environment.

The rise in the number of people travelling by air, coupled with the changing social profile of the passenger market, has since the mid 1980s provided airports with a series of new commercial opportunities. Airports now have the volume, range of passenger profiles and characteristics to mirror domestic retail markets much more closely. The growth of branded specialist retail chains with their package of uniformity, high-quality standards, franchise management and, in many cases, international branding, was designed for easy transfer to any location. The airport environment remains a logical area for expansion, offering retailers a defined customer base with a higher than average level of discretionary spending power. Most of these retail chains operate out of small, customised shop units requiring relatively little backroom storage and working area. As a result they are easily adaptable to an airport setting.

It is often said that commercial managers within airports will take all available space within a terminal in order to maximise non-aeronautical revenues. Gray (1997) suggests that such a strategy would ultimately have a negative effect as the commercial strategy would be achieved at the expense of other operational functions. There remains however no precise formulas to determine the correct balance between commercial relative to non-commercial space within an airport. The challenge for airport management is to optimise the non-aeronautical revenue of the terminal without compromising operational effectiveness.

The determination of commercial space requirements will be influenced by a variety of factors. These include the composition and number of travellers, the number and type of flights (i.e. long-haul or short-haul; intercontinental, European or domestic; leisure or business; scheduled or charter) and whether the airport follows a commercial or traditional development strategy (Doganis, 1992). This latter factor is fundamental to the optimisation of non-aeronautical income and remains a highly controversial issue between the aviation and commercial managers of airports.

Despite the growing requirement for airports to develop alternative sources of revenue, many managers on the aeronautical side continue to view commercial activities with suspicion. They believe that the deliberate placement of retailing in the direct path of the passenger is detrimental to the achievement of the airport's primary function, which is to process travellers efficiently, safely, on time and to the right aircraft. Some airport authorities such as BAA have extended and adapted terminal buildings to allow for additional retail outlets in order to avoid compromising core aviation areas and services through retail development. This approach has allowed a much greater density of retailing in the extended terminals but has also fuelled controversy as to the right balance between the retail offer and the ease with which passengers can access the aircraft departure gates. As airport capacity grows, the walking distances needed to reach departure gates is lengthening and becoming a matter of some concern to airport authorities.

The demand for airport retailing comes from two sources. First, passengers enjoy the experience of shopping and consider it to be an integral part of air-travel. Secondly, airport authorities have come to rely upon the revenue it provides and support the continued provision of additional airport facilities. The importance of commercial activities is underlined when airport authorities and their shareholders contemplate the size of many proposed investments. Dusseldorf, for example, is estimated to need over DM1 billion for redevelopment.

Architectural Planning

The requirement for retail activities within an airport highlights a further controversy between architectural design and commercial planning. Even before attention is turned to the retail tenant mix within an airport or the precise location of individual retailers, the debate between commercial priority and overall design aesthetics has to be considered.

While revenue from retail activity has become increasingly important, it would be misleading to suggest that the transition toward a more commercially orientated development has been unproblematic. Extensive consultations involving the aeronautical, commercial and architectural interests of the airport have been a prerequisite to retail development.

Some such as Gray (1994) maintain that few architects have been sympathetic to the need to maximise returns from commercial revenue. It is argued that the typical design for a terminal is architecturally driven. Airport authorities in commissioning internationally known architects have in the past failed to represent the interests of different groups within the airport. Operations, maintenance and especially the commercial sector need to be properly represented in the design and building process. Controversially it could be argued that in some instances this has not happened and the airport authorities have abrogated any responsibility for the project. A consequence of this approach is to place significant power in the hands of the architect over alterations, adjustments and final planning decisions.

The design of an airport terminal is a highly complex undertaking with the requirements of many parties having to be accommodated in a fully integrated building. Traditionally the airside/landside configuration of a new terminal layout has been based on the positioning of the main building service cores and the primary airport services such as check-in, baggage handling and security/'friskem'. Once the positioning of these functions is decided, the flow lines for the airport's passenger traffic are in effect fixed. Trying to 'tack-on' retailing facilities to such a complex structure inevitably means that the terminal will never be utilised to its maximum commercial potential. Consequently the airport authority will lose significant revenues over the life of the terminal investment, which is normally considered to be in excess of forty or fifty years (Gray, 1997).

In these situations not only will there be insufficient retail space to meet demand, but what is provided is often badly located and has been added as a secondary consideration. The process of planning and design has therefore been characterised by a power relationship in which the airport's commercial operations have been subservient to the demands of other parties.

To achieve the necessary integration of retailing and other commercial activities into the terminal process, Gray (1994) maintains that the retail planner needs to be involved with the architect from the initial meeting. The role of the planner will be to 'challenge the allocation of facilities and to achieve the best possible compromise between operational efficiency and maximum revenue from retail property' (p. 2).

Fuelling the argument for this development have been changes in consumer expectations in relation to the environment and product and service standards. A high-quality retail offer has become a feature of many towns and cities throughout Europe. Customer-friendly layouts complement high design standards and consumers now expect the very best facilities wherever they decide to shop.

Walsh (1992) argues that the traditional view of the planning environment within airports was to create differentiation through grand architectural statements or expensive marble-lined interiors. Today such an approach, unless it is also 'humanised', i.e. it includes people-friendly processes and facilities, is unlikely to appeal to the consumer. The aim for the airport authority is therefore to develop a corporate policy towards airport and terminal design and one that establishes a clear differentiated market position. In so doing there is a need to ensure that the passengers' interests remain paramount and that aviation and commercial requirements are fully integrated into whatever design concepts are accepted.

Evidence suggests that attitudes have begun to change in the planning community. A greater willingness to design a terminal with commercial activities in mind has been noted. Retail design practices are increasingly in demand by airport authorities anxious to optimise the financial return from investments in terminal buildings and facilities. Aer Rianta's new terminal extension and pier at Dublin airport, for example, has been designed from the initial concept stage with the input and approval of the airport's retail and commercial management. BAA have gone a stage further and used their own commercial management to plan the retail space requirements in terminal extensions with little architectural involvement. Furthermore, some retailers such as the Swiss-based Weitnauer Group have actually assisted in the redesign of some airports where they have held commercial concessions.

The Planning Process

The proportion of revenue now derived from non-aeronautical sources has resulted in commercial departments becoming increasingly influential. The recruitment of professional retail managers from outside the airport sector and the existence of a commercial director on the main board of many airport authorities have further consolidated the power base of non-aeronautical interests. While the decision to expand commercial activities still cannot be achieved without considerable internal controversy, the base from which negotiations begin has been considerably strengthened.

With greater representation at a senior level, commercial managers now talk in terms of 'an integrated approach to airport design' and one that takes a 'seamless approach to passenger facilitation'. New terminal development requires the involvement of airport management at the most senior level. This process remains complex and is often divided into a series of distinct stages.

The first stage will require agreement on the precise location of the new terminal. In most cases the airport's Master Development Plan will have already dictated the location of a development on the scale of an airport terminal. Master planning within the airport is essential in order to maintain an orderly process on the development site. It ensures that the necessary lands are available and that all construction complies with the safety and operational requirements vis à vis the aircraft apron and airfield areas. A further important consideration is that no single project should put constraints on either the existing airport operations or future infrastructure developments.

The second stage is a data-gathering exercise that requires input from a number of different sources. These may be internal to the airport authority or external, and include airline and aviation traffic consultants. Typical information required would be passenger traffic forecasts and expected customer profiles. The forecasts, which cover short-, medium- and long-term traffic levels for up to twenty years, provide the basis for setting the capacity parameters of the proposed development. In addition, management must identify the likely aircraft fleet composition of the user airlines and create the appropriate service and space standards for the new terminal.

On the basis of these discussions the third stage of development requires an expert planning and operations group from the airport to draft a detailed briefing document. This is then presented to the project's architectural and engineering consultants. The briefing document forms the reference base for all subsequent design solutions and sets out both the macro and micro level parameters that the design should meet. A briefing document would typically contain information on:

- the number of passengers to be handled in a typical peak hour;
- capacity requirements for the anticipated number and type of aircraft;
- the level of flexibility required to cope with changes in aircraft design and type;

- requirements for future expansion;
- the aircraft-servicing facilities required.

In addition, the facilities considered necessary in the terminal are listed in detail along with any specifications for cost-effective energy systems such as solar lighting or low building maintenance costs.

Depending upon the size of the airport operator and the scale of the new development, an airport may choose to use its own in-house consulting facilities or contract out to third party designers. In the fourth stage of the development process, the designers using the briefing document create the initial plans for the new facility. These plans will indicate the amount of space required for both commercial and aeronautical activities, provide drafts layouts for the terminal, and identify the location of the principal passenger and airline facilities (including retailing). It is at this stage that the broad passenger flows are fixed, and arriving and departing passengers segregated. The division between the landside and airside areas is often the subject of considerable discussion. Countries that are party to the Schengen agreement (which includes all members of the EU except the UK and Ireland) have no border controls between member states. Consequently they are required to separate Schengen from non-Schengen passengers. This has considerable implications for terminal design and leads to a replication of facilities and services for each group of travellers.

In fixing the landside/airside divide, the airport operator is required to have a clearly defined retailing policy that effectively targets departing, arriving and non-travelling customer segments. Correct market positioning should separate the retail offer between these groups and ensure that a cannibalisation of sales does not occur. For example in Paris CDG 2 airport, the landside shopping facilities are located on the arrivals level. Departing passengers are therefore unlikely to shop at these facilities and will undertake the majority of their purchases airside. While landside shopping is becoming increasingly popular, the majority of airport authorities in Europe have focused upon airside retailing. Planning policies have therefore concentrated upon the development of duty/tax-free speciality shopping.

Once the landside/airside divide is fixed, agreement is reached on the sequence and location of security and passport control positions. The remaining airside space within the terminal is available for retailing and other commercial activities. Making the correct set of allocatory decisions at this stage is important, as having to redefine usage when the terminal is fully operational will be an expensive and disruptive process.

By this stage the budgetary implications of the brief will have begun to emerge and an initial estimate of the likely cost of the project will be possible. On the basis of these costings there may be a requirement to scale down some of the airport's requirements. This can result in extensive and protracted negotiations between the various interested parties in the airport. It is the involvement of commercial management at these initial stages of the process that represents the most radical departure from the traditional approach to airport planning. By ensuring a commercial input to the briefing document itself, factors such as legislative change and emerging retail trends can be incorporated into any initial calculations on commercial space requirements.

Similarly, the factors that differentiate airport retailing from domestic retailing can be identified and accommodated within the design. Both the airport passenger-handling process itself and the flight departure times dictate that a customer has limited time in which to shop. Such influences have obvious implications for retail location, layout and the shopping process itself. In addition, all duty/tax-free shops are subject to Customs and Excise controls, a complication that high street retailers do not have to contend with. A further consideration from a terminal-planning perspective is whether to situate the retail warehousing (and in particular, the bonded warehousing required for duty-free product) within the airport terminal or at a remote location. The merchandise supply routes to these shops from the warehouses will also have to be integrated into the terminal design and will be subject to Customs and Excise approval.

Once senior airport management has approved these initial plans, the next stage is to draw up a much more detailed design package. Close monitoring of the emerging costs is essential at this stage. Regular liaison between the airport management and the project consulting team (which by now will include quantity surveyors and structural engineers) will enable any specification adjustment to be made early enough in the process to avoid budgetary implications. A full architectural and engineering team will be responsible for developing detailed models, drawings and computer-simulated walk through video demonstrations of the new facility. The objective will be to ensure that the airport operator has a clear understanding of what is being developed. Once the construction of the terminal begins then the airport authorities assign a project planning group to oversee developments. Again the importance of non-aeronautical functions are in evidence as many operators now have a member of the commercial department represented on this planning team.

The decision on the precise location of retailing facilities and the allocation of space between, for example, the duty-free sector and the more general tax-free outlets is a matter for joint decision making between the airport commercial management and the project consultants. Within the consultancy team a retail planner will ensure that each individual outlet is located sequentially in a way that allows maximum exposure to the international passenger flows and achieves optimum customer spend for the airport operator.

Calculating Retail Space

Airports tend to work to a series of specific formulas in order to determine their optimum retail space requirements. Too little space given to an outlet will result in the unit becoming overly congested, creating queues at the check-outs. This may lead to dissatisfaction and because of the time constraints on passengers, result in customers choosing not to make a purchase. Alternatively, too much retail space will result in poor sales per square metre and an opportunity loss for the airport authority. Both scenarios lead to a less than optimum return.

An airport operator will therefore need to determine the level of retail space that each passenger should be entitled to within a store, identify the location of the outlet (e.g. departures lounge/gate lounge) before calculating the area required for storage and staff facilities. A traditional view was to allocate one square foot of retail floorspace for every 300 passengers travelling per annum or one 12,000–13,000 sq. ft outlet for every 400,000 travellers. Apart from this general rule of thumb, a wide variety of different methods are employed to determine the optimum size for a trading unit. What is provided below is an example of one of the more simplistic methods used within the airport industry.

For example, to determine the sales area for a duty-free store in the departure lounge, both the space per passenger and the number of passengers present need to be calculated. Therefore

$$\text{Space per passenger (m}^2) = 1 \text{ (personal space)} + 2 \text{ (circulation space)} + 1 \text{ (tills/gondolas)}$$
$$= 4\,\text{m}^2$$

The area of sales floor (A) required is therefore calculated as

$$A = 4 \times \text{N}$$

where *N* is the number of passengers present, calculated as

$$N = T/60 \times F \times TPV \times DFR$$

where

T = average dwell time in duty-free shops (minutes, e.g. 14);
F = peaking factor (suggested value 1.2);
TPV = target penetration for visits (suggested value 0.65);
DFR = design flow rate.

Therefore for a terminal with 3200 passengers/hour

$$N = 14/60 \times 1.2 \times 0.65 \times 3200$$
$$N = 582$$

and

$$A = 4 \times 582$$
$$A = 2328 \, \text{m}^2$$

For satellite and gate lounge shops the calculation will be similar, however the penetration level will be lower, e.g. 0.25, as will the dwell time, e.g. 5 minutes. The equation may therefore be rewritten as

$$N = T/60 \times F \times TPV \times S \times DFR$$

where S = the proportion of passengers passing through the satellite (e.g. 0.3), giving

$$N = 5/60 \times 1.2 \times 0.25 \times 0.3 \times 3200$$
$$N = 24$$

therefore

$$A = 4 \times 24$$
$$A = 96 \, \text{m}^2$$

The area required for storage will depend upon how the company operates its supply chain and the rental rate it has been able to negotiate with the airport authority. Many retailers find the cost of hiring storage space within an airport prohibitively expensive. For example, in Schiphol, KLM who operate the airside alcohol and tobacco concession has a warehouse approximately 8 kilometres away from the airport. The concessions need to be restocked every two hours and the company makes ten deliveries per day. This however is estimated to be more cost effective than renting space at the airport, which is calculated to be between eight and ten times more expensive than equivalent space in the local environs.

The area required for storage is therefore highly variable and is calculated as a percentage of the total floor area for the store, i.e.

$$W = A \times X$$

where

W = area required for storage;
A = area of sales floor required;
X = estimated percentage of sales floor area required for stock (e.g. 10%).

Therefore

$$W = 2328 \times 0.10$$
$$W = 233\,\text{m}^2$$

Determining the level of space for a staff area requires an estimation of the total number of employees that will be working in the outlet. Typical requirements would be a social space and a locker. If the assumption is made that 20% of employees would be in the staff area at one time and all staff would require a locker, then we are able to calculate the following:

$$E = (L + (Sn \times As))$$

where

E = staff space per employee required;
L = locker space (e.g. $0.5\,\text{m}^2$ per person);
Sn = staff present in social space at one time (e.g. 20%);
As = access space required per person (e.g. $1.5\,\text{m}^2$).

Therefore the space per employee is calculated as

$$E = (0.5 + (0.2 \times 1.5))$$
$$E = 0.8 \, \text{m}^2$$

If the number of staff required is measured as 65 for each 1000 passenger hours and the design flow rate is 3200, the total floorspace for staff facilities should be

$$TSS = SR \times DFR/1000 \times E$$

where

TSS = total staff space required;
SR = staff required.

Thus

$$TSS = 65 \times 3200/1000 \times 0.8$$
$$TSS = 166 \, \text{m}^2$$

Tenant Mix

Many airports attempt to project a clear identifiable image in the eyes of the consumer in order to differentiate themselves from the competition. The retail offer represents a key method of achieving this strategy. Currently many airports attempt to create the atmosphere of a shopping centre by having stores that are either well known on the local market or established internationally. In this way airport customers can identify with many of the brands available.

While the emergence of branded specialist retailers has facilitated retail differentiation in many airports, not all operators have followed such a strategy. Some airport authorities provide a much wider and specialised range of product offerings in outlets with generic names such as 'fashion', 'electronics', 'cameras' and 'confectionery'. Schiphol and Singapore are examples of airports that have opted for unbranded generic retailing.

The way in which an airport operator can seek to influence its commercial revenues is through a co-ordinated and well-planned retail tenant mix. Chesterton (1994) notes that Britain's 15 largest airports had a total of 248 speciality shops occupying a total of 323,792 square feet of retail floorspace. This is expected to increase to over 600,000 square feet by the end of the decade (Shaw, 1993). Schiphol airport has a mix of retail offers currently selling a range of over 120,000 different products. The new Chek Lap Kok airport at Hong Kong will have 150 retail outlets, while in the Middle East, Abu Dhabi has provided 400 sq. m for a second level of specialist shops and will allocate 4000 sq. m to retail in its new satellite terminal.

Managing the airside and landside retail facilities within an airport has been likened to operating an out-of-town shopping centre. The need to manage the tenant mix, to provide support and maintenance and to market the centre has distinct parallels. Shaw (1993), however, maintains that airport operators keep tighter control over their tenants than shopping centres by stipulating opening hours, product range, security conditions and standards of cleanliness within the store. This reflects a number of unique differences that distinguish airport retail areas from domestic shopping centres. For example, Shaw (1993) maintains:

- Airport terminals are primarily to process airline passengers efficiently and safely, retailing is not the prime purpose but a secondary activity.
- There is a greater predictability of demand within an airport. For airside retailers especially, the available population can be defined precisely as to numbers, destination, purpose of travel and profile.
- The customer market is international by nature and travellers tend to be in the higher income bracket.
- Typically passengers who are about to depart or are in transit have higher disposable and discretionary income and a greater propensity to spend.
- Passengers have a limited amount of time and little opportunity to gain familiarity with the store or to return goods.
- The majority of airports are open every day of the year and in many cases twenty four hours a day.
- Many airports shops are required to accept a wide range of currencies at point of sale.
- Customs and Excise quantity and value allowances apply in airport duty/tax-free shops.
- Airport duty/tax-free retailers are subject to a range of Customs and Excise accounting and warehousing controls and regulations.
- Opportunities to build a loyal customer base remain limited owing to the low travel frequency of most passengers.

While such factors have a range of marketing and HRM implications, their importance in this context is in the influence such criteria have over the tenant mix for the airport. Popular brands such as Bally, Levi, Body Shop, Tie Rack and Sunglass Hut have become permanent, established features in many airports. Dissonance related to the purchase decision can be significantly reduced by the inclusion within the tenant mix of these branded retailers. The instant recognition of branded products can lead to shorter purchase decision times, which may allow time-limited passengers the opportunity to make incremental purchases.

A second factor influencing an airport authority's choice of tenant mix is the need to develop a contingency plan in respect of 1999. As noted in the previous chapter, the abolition of duty free represents a significant loss of revenue for the airport authorities. An eclectic array of retail specialists would help to offset the potential impact from this event.

In addition to attracting specialists and international branded retailers, the objective has been to use concessionaires to provide a flavour of the city or region that the airport serves (Klapper, 1995; Bingman, 1996). The merchandise in the shops of the Museum Company includes artefacts that are on show in local galleries and museums. Schiphol has an outlet specialising in the sale of flower bulbs, while Edinburgh has a concession selling Scottish salmon and haggis, and Milan Linate has significant sales of Chianti, olive oil and Parmesan cheese. Specialist food shops have emerged in many airports on both landside and airside areas and offer time-pressed travellers the opportunity to purchase local produce. These products are often packaged as quality gifts and represent a convenient means of using any remaining local currency.

While the majority of airport retail concessions are awarded on the basis of a tendering process, the airport operator is required to be proactive in managing the tenant mix. For larger airports with a range of different sized retail units, the objective may be to develop a balanced portfolio of known branded retailers. Specific companies may be identified and approached by the airport operator. The tendering process (described in Chapter 2) may be by-passed and a retailer requested to consider occupying an available site.

A strategy in the larger airports may also be to include a wider range of luxury and peripheral retail units that appeal to only a minority of passengers. Once an airport's traffic grows above seven million it reaches the critical mass needed to sustain a wider range of niche retail outlets. For example, antique dealers, carpet and tapestry traders, high-quality jewellers and lace makers may be found in some of the larger airports. Despite appealing to only a small segment of the travelling population, large passenger volumes have made such specialised retailing a sustainable proposition.

Similarly as airports have grown, increasing numbers of non-travelling customers, including large numbers of people working at the airport, have utilised the landside retail facilities. Larger airports can have tens of thousands of people employed on site. Airport operators have sought to meet this demand through the provision of retail outlets specifically targeted at the non-traveller. For example, Heathrow has a concourse of stores away from the main passenger thoroughfares that includes a newsagent, a convenience store and a photo-processing centre. Other airports have food and general supermarkets, laundry and cleaning outlets, travel agents, pharmacies and post offices.

For smaller airports with limited passenger traffic, creating such diversity will not be an option. The operator will focus upon establishing a core retail offer with non-overlapping assortments based around duty/tax-free items, books and news, and catering. While retailers may be expected to tender for the duty/tax-free concession, the airport authority may have more difficulty in attracting other retailers to operate within the terminal. In this instance some companies, such as Alpha Retail Trading, may offer a complete package to an airport operator. If successful in the tender process for tax and duty free, they may operate other fascias from their retail portfolio such as a book shop (BooksPlus) and a drugstore (Drugstore). From the airport's perspective such an offer remains attractive, as it allows the airport to negotiate with a single concessionaire while at the same time providing a wider retail offer.

The matching of the tenant mix to consumer demand is illustrated in a number of airports. At Heathrow, the retail composition for each terminal is a reflection of the passenger profile. For example, Humphries (1996) notes how half the passengers at Heathrow Terminal 1 are on business and well over half are men. Consequently branded products are sold through established quality retailers such as Austin Reed, Thomas Pink and Links of London. In Terminal 2 the high proportion of European passengers means that there is considerable demand for traditional English goods. Passengers using Heathrow Terminals 3 and 4 are on flights to Scandinavia, the USA and the Far East, and have above average levels of disposable income. Their propensity to purchase high ticket goods is consequently higher. The retail tenant mix therefore reflects this with a variety of branded goods being sold through established retailers.

In contrast, Manchester airport has a higher proportion of charter flights and consequently the number of retailers selling high-quality, branded merchandise is reduced in favour of mid-range clothing, music and accessories, for example Dixons, Our Price, Dorothy Perkins, Burtons, Warner Bros. and Tie Rack.

The composition of the tenant mix within a terminal may also change to reflect a repositioning of the airport. This is illustrated at Gatwick airport, which traditionally had a high proportion of chartered flights. As British Airways has developed its operations at Gatwick, the traffic profile is becoming more typical of a main hub airport with a mixture of business, scheduled and charter flights. The retail composition, particularly in the newer north terminal, attempts to reflect this changing market.

To ensure that sales revenue from retailing is maximised, a number of airport operators have adopted a proactive approach to managing the tenant mix. Detailed passenger data, traffic forecasts and route profiles are provided to retail operators to help in their marketing activities. Regular review meetings, seminars and even 'away days' are arranged to analyse in detail the business performance and jointly plan for the period ahead. Many airport authorities are organising joint promotional campaigns in co-operation with the concessionaires and the leading merchandise suppliers (see Chapter 4).

Tenant Location

Few people could have predicted either the volume or the speed by which airport passenger numbers have grown over the past two decades. While on the one hand this has represented a welcome revenue stream for operators, many airports find themselves constrained by the physical size of the terminal with limited, or even no space to expand further. Emphasis has therefore to be placed on ensuring that retailers are located in areas that maximise consumer spend and give the best return to the airport.

Doganis (1992) maintains that retail facilities should be located in the direct line of the passenger flow and as close to the departure gates as possible. Because many travellers experience a degree of anxiety when travelling by air, few will deviate from the main passenger routes in order to purchase retail products. The airport passenger processing system creates its own momentum as individuals are moved from one stage to the next (Figure 3.1).

From the check-in desk, passengers are directed to the departure gates via the passport control, security checks and departure lounges. Flight information screens and PA announcements continually encourage the movement of passengers towards the embarking aircraft. One

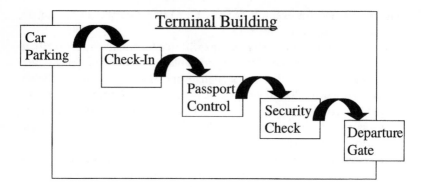

Figure 3.1   *Airport passenger processing system.*

implication of this for retailing is that the process creates a dynamic of its own and, unless retail outlets are situated close to or in the actual passenger flow line, only the most highly motivated passengers will search for the shopping area. Such a situation was illustrated at Brussels International airport where a two-storey retail shopping centre had vacant units on its higher level and had difficulty encouraging a move upstairs from the high-quality shopping offer on the ground floor.

In identifying the premium store locations on the airside of a terminal, there remains some debate over how the retail offer should be configured. Bingman (1996) sees this as a dilemma for the airport operator. If concentrated in one specific part of the departure area, it creates the visual appeal of a shopping centre, which in turn creates synergy between outlets and increases the propensity of consumers to spend. Alternatively, if the retail offer is spread out across the whole of the departures area, it provides the customer with a greater number of opportunities to purchase products. Airports such as Copenhagen have gone for the former approach while Schiphol has two shopping areas, both identical, existing within the departure complex.

In identifying the key factors influencing the ideal retail location within an airport, four criteria are therefore important:

- *The logic of the passenger traffic flows:* the location of shop units needs to mirror the direction in which the passengers are travelling. They should not be required to retrace their steps or go in a counter direction.
- *Floor levels:* retail outlets should be on the same floor level as the departure gates and passengers should not have to ascend or descend stairs in order to shop.

- *Distance*: shops should be accessible without passengers having to traverse long distances. They should be sufficiently removed from the security and passport checks to allow the traveller to make the mental adjustment to a shopping environment.
- *Visibility*: before encountering the retail offer, passengers should have the retail outlets in their line of vision. This will help stimulate purchasing behaviour and possibly trigger impulse sales.

Store Design and Layout

Once the basic space requirements have been calculated for a trading unit, the objective is to design a store that is consistent with the airport's overall market positioning and passenger expectation. For concession-aires, customer-related information, such as passenger mix, airport penetration levels and dwell times, will have been considered in the tendering process. There remain many parallels between the design and layout of stores in an airport and a high street. Both have to ensure that the layout encourages customer flow, that the visual display of the store reinforces the marketing image, that the merchandise can be located with ease and that there are adequate security and storage facilities.

Despite these similarities there also remain some notable differences between the airport and high street retail outlets. These differences are significant enough to influence the way in which such stores are de-signed, laid out and merchandised. The airport retailer must consider a number of factors in order to ensure that the store remains both aesthet-ically pleasing and functionally efficient. Among the most important fac-tors that an airport retailer must consider, are:

- access to store;
- layout and circulation within the store;
- pay points;
- design and aesthetics;
- equipment and services;
- merchandising of product.

Access to store

The objective for the retailer is to make store access as easy as possible for the consumer and to encourage maximum passenger entry. For air-side retailing this requirement raises a series of specific issues. Unlike many high street stores, the opportunity to utilise large visual window displays to attract customers remains limited. The reason for this is two-fold. First, given the premium rates that are attached to airport floor-space, few retailers can afford large areas of external display. Secondly, as passengers have time constraints, retailers do not wish to encourage potential customers to remain outside the store as this reduces the amount of in-store browsing that is possible.

One method used by airport retailers to stimulate access into their shops is therefore to make the entrance area as wide as possible and to have no physical barriers to entry. Such a strategy has its disadvantages however. The majority of passengers carry some form of baggage on their person throughout their journey, including the period when they are shopping in the airport. This may mean circulating with their baggage carts in the store. From a retailer's perspective, allowing baggage trolleys into a store increases the amount of in-store circulation space required and reduces the areas that can be devoted to displaying product.

An open entrance into a store is also thought to encourage shrinkage. A number of retail outlets have therefore erected one-way entrance gates or turnstiles. This has the potential to cause difficulties for customers. Many are unable to bring their luggage into the shop because the trolley either does not fit through the entrance or the retailer has banned trolleys from the store. Moreover, they are prevented by the airport authority from leaving their baggage unattended outside the shop for security reasons. Access for the disabled passenger is another important consideration in creating entrances into a store. For example, where changes in floor levels are necessary, ramps must be provided.

Layout and circulation within the store

The two basic rules in planning the layout of the sales floor are to use all available space and to balance function against aesthetics. Achieving the correct layout remains important as it will encourage customers to visit all merchandise areas within the store, while at the same time allowing particular sections to be highlighted. It can also reduce the number of unproductive and obscured areas and provide greater merchandise security. Furthermore, a successful layout provides better control over the product range by allowing incompatible products to be separated and seasonal displays to be featured (Barr and Broudy, 1986).

The layout of an airport shop will depend upon a variety of factors including the size and format of the unit, the type of merchandise sold and whether it is a self-service or counter-service operation. In general, the merchandise on sale at airport retail outlets falls into two categories, consumables and non-consumables. Duty-free outlets as a rule concentrate on selling consumable merchandise including liquor, wines, beers, tobacco and confectionery products and, as with domestic grocery retailing, use the 'supermarket' format. The sale of non-consumable products on the other hand tends to follow a 'boutique' or department store layout.

As already discussed, airport retail customers are under time constraints and therefore the speed of transaction is an important factor in the maximisation of sales. This has led to the majority of airport retailers opting for a self-service layout. Perfume and cosmetic products, for example, which are traditionally sold across the counter in the high street, are increasingly available on a self-service basis (or 'self-assist' as termed by the fragrance houses).

The self-service approach was strongly resisted by the fragrance suppliers for many years as it portrayed an image contrary to the one they wished to develop for their products. Airport retailers, however, faced with ever increasing passenger volumes and the quickening pace of new product introductions, began to experiment with the self-service concept. Between 1996 and 1999 over one hundred new women's fragrances will be launched by suppliers to the duty-free market (DFNI, 1997). Aldeasa, the Spanish duty-free operator, was one of the first companies to introduce a completely self-service based perfume store at Madrid's Barracas airport in 1995. The problem of pilferage in a self-service store remains of concern to many operators, given the relatively high unit value of perfume and its easy portability. However, by positioning perfume consultants (representatives of perfume companies) and sales assistants at the free-standing merchandising units to offer advice and help to customers, sales increases from self-service have outpaced shrinkage by a factor of ten.

The basic envelope of the retail unit is fundamental to the layout decision. Two types predominate. First, the 'corridor' envelope, in which customers enter at one end of the store and exit at the other, allows the retailer considerable layout flexibility. Secondly, the 'switch-back' envelope has one opening, through which customers must enter and exit, thus requiring a more complex layout (Figure 3.2). The retailer normally has no choice with regard to the outlet envelope, as it is most often dictated by the configuration of the building itself which, if designed without a retail specification, is often unsympathetic to consumer needs.

Airport retail units are small-scale by general retail standards, owing in part to the overall capacity constraints of many airport buildings and to the high cost of providing commercial space. The important size

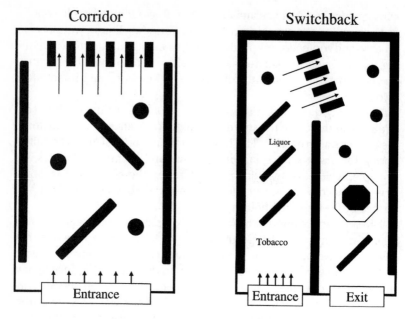

Figure 3.2   *Airport retail unit envelopes.*

variables in the airport retail unit relate to the width and depth of the outlet. Too narrow and the layout becomes claustrophobic; too deep and passengers may be reluctant to enter.

Retail layouts within an airport are required to have a logic that the potential customer can quickly comprehend and assimilate. A clear route into and through the store must be easily identifiable in order to reassure passengers that their departure gate remains accessible at all times. Equally, passengers should be able to decide if an outlet is of interest to them without having to spend time trying to understand what the store has to offer.

A store layout, in order to achieve its objective of ensuring the optimum presentation of merchandise, needs to be proactive. Layouts that are described as 'passive' are those that leave the customer to follow a totally random route through the store or allow a direct line to the checkout (Figure 3.3). In the airport retail environment where time is so important and random browsing is uncommon, layouts need to manoeuvre or steer the customer to as many merchandise sectors as possible, in particular those with the higher margins. Examples of proactive layouts are the 'snake' and the 'pin-ball' (Figure 3.4).

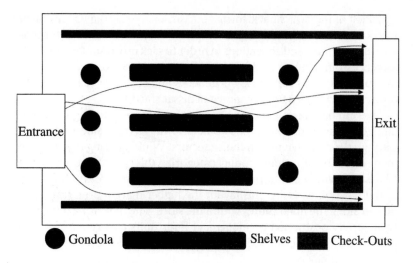

Figure 3.3 *Customer flows in a passive airport retail layout.*

The 'snake' literally directs the customer through the store and past all the merchandise. The customer has no choice but to follow the route provided or go back to the entrance. This layout will work in either the

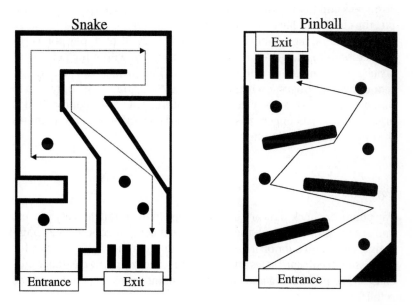

Figure 3.4 *Proactive retail layouts.*

'corridor' or the 'switchback' format but can antagonise regular customers who know what they want to purchase and where to find it. Having to go past all the merchandise sections in order to pick up one or two items can affect customer relations. If the circulation corridor is too narrow, popular product areas can cause bottlenecks and customers can be slowed down which will be a further source of dissatisfaction.

The 'pin-ball' layout uses the positioning of free-standing floor units and gondolas to encourage customers to move from side to side and through the different merchandise sectors. While steering the customer in the direction of various product categories, this type of layout provides passengers with the option of quickly moving to other areas of interest or to the pay points. Using this layout the higher margin products can be placed along the main routes within the store. The need for a number of strategically positioned gondolas and floor units also requires a store envelope with sufficient width and depth to allow the necessary passenger circulation.

In smaller airports, where the space available for retailing on the airside is limited and passenger volumes are small, both duty-free and tax-free shopping are usually combined in the one location and are operated by a single retailer. This situation requires a multi-format approach, using a supermarket style layout for the sale of consumable duty-free goods and a general department store format for the sale of perfume, cosmetics and tax-free merchandise. In Birmingham airport, for example, the checkouts have been positioned between the supermarket stage and the general merchandise area, while in Cork airport, pay points are located both in the general merchandise section with a further line of checkouts at the store exit (Figure 3.5). The former can be used to reduce overall congestion in the store by allowing those who only wish to purchase duty free a fast exit from the store. The latter approach, it is argued, draws passengers past all the merchandise areas and increases the opportunities for impulse purchasing.

Pay points

Pay points or check-outs are the final stage in the retail delivery process and are the point at which the transaction is completed. Each pay point consists of a unit supporting a cash register, together with facilities for processing credit card sales and for packing and wrapping customer purchases. The positioning and type of cash point varies, depending on the retail format.

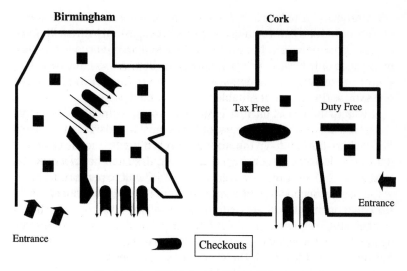

Figure 3.5 *Checkout locations within stores.*

Airport retailers using the supermarket format to sell duty-free consumable products require the check-outs to be located in one group, usually just before the store's exit point. The number of check-outs required is a function of the forecasted passenger penetration rate at the **peak** hour (i.e. the number of passengers present in the store during the busiest hour of the day) and the time required per transaction. The transaction time calculation should also include an allowance for the number of queuing customers that the retailer is willing to accept at any one point in time. An important consideration in determining queue length is the time constraints upon customers and the low tolerance to any form of delay.

The check-out unit must be ergonomically designed to ensure that the cashier can work in comfort, accurately record the products sold, operate the cash register, process credit card transactions and pack goods. The units must allow sufficient space for the customer's purchases to be processed safely and efficiently, and provide for the storage of packing materials. Ideally, the register system ought to be a fully integrated EPOS package which links all the check-out points with the cash office, the warehouse and the product buyers. Laser-driven bar code units are now the standard in many airports, with the scanning devices being incorporated into the check-out unit itself.

The configuration of the check-out units remains a subject of debate. Space availability and the desire to maximise the floor area for merchandise and selling are important considerations in how the check-outs are arranged. Sufficient area has to be provided in front of the check-outs to allow for some queuing and to avoid congestion in the selling areas. The space between check-outs can be less than required in domestic markets as only baskets or small narrow trolleys suitable for hand baggage are provided to passengers. As with access to the store itself, at least one extra wide check-out suitable for the disabled passenger is essential.

It is equally important that customers having completed their purchases are able to leave the check-out area quickly and not cause congestion. Where the exit from the store is narrow it is possible to stagger the line of check-outs in order to achieve the required number of pay points. With this configuration, however, it can be difficult to persuade customers to use the second line of check-outs and it may be necessary for the store management to direct customers to available tills.

The most frequently used configuration has one line of check-outs which follows the exit line of the store. When using the multi-format layout discussed above, with the check-outs positioned at the cross-over point between formats, the need for adequate circulation space becomes critical. Queuing in the middle of the store needs to be avoided and customers allowed to make the transition from one format to the other.

Pay points for the boutique and department store formats can either be stand-alone units or incorporated into the run of the counter. From a customer service perspective, the greatest flexibility is achieved when all pay points are fully integrated into the EPOS system and can take payment for any type of merchandise. This facility not only speeds sales processing time by reducing the number of transactions a customer has to make but also helps to avoid queuing at the most popular merchandise areas. As in the case of check-outs, these pay point units must allow for the safe processing and handling of the goods and provide storage space for packing materials.

Design and aesthetics

Design and aesthetics remain important factors in determining the success of an airport's retail operation. Creating unique, attractive and exciting retail environments is essential if stores are to maximise passenger penetration levels. Airport customers are the same individuals who shop in today's high-quality, city arcades and out-of-town retail centres. Such consumers are increasingly discerning and have come to expect a retail offer of similar quality at the airport. Many of Europe's airports have already developed new terminal buildings, with many more in the course of construction or refurbishment. New capacity is finished to a very high specification and airport authorities expect the retail outlets in these terminals to be of a similar standard. The design of airport retail outlets should ideally reflect national architectural motifs and styles and involve quality finishes and surfaces, preferably using local materials. The use of marble, stone, brick and high-quality timbers will contribute to the creation of quality retail surroundings that complement the terminal building and the branded products sold in the terminal's shops. The overall ambience created in-store must be integrated with the merchandise presentation and should aim to reflect the tone, character and style of the goods on sale. This combination of in-store design and display with a co-ordinated merchandise strategy is an essential component in the market positioning of the store.

By blending local materials and styles with modern fittings and lighting design, it is possible to design and develop an airport's shopping offer to an international standard yet create a sense of local culture and place. Such an approach will help to differentiate the airport in the mind of the passenger and identify it as somewhere special and memorable to shop. Schiphol, Copenhagen and Gatwick South are examples of airports that have achieved a high degree of differentiation by using this approach.

Design and aesthetics can be used to integrate the retailing areas with the terminal's passenger lounge in order to create a relaxing and inviting environment for the airport user. Displaying similar design features and graphics throughout the terminal and the use of common floor, wall and ceiling finishes have the potential to integrate a diverse range of shopping outlets into a cohesive retail centre.

Aside from providing pleasing aesthetics, airport retail design also has to be practical. Many millions of passengers pass through airport terminals annually, resulting in the retail facilities being in use every day of the year. Quality, durability and easy maintenance are important characteristics when developing a retail design specification. For example, this is particularly important in relation to floor surfaces in the duty-free outlets. As breakages and spillages are common in these outlets, the floor surfaces will have high non-absorption, stain resistance and non-slippage characteristics. Combinations of hard and soft flooring materials are frequently used to create tactile variety and to provide a more durable surface for the main customer flows. As in downtown stores, flooring materials are also used to direct customers through the store and past the principal merchandise groups.

In developing the retail outlet, the designer must provide a quality environment not only for passengers but also for employees. Effective air conditioning that creates a comfortable environment for both the shopper and the staff, while at the same time maintaining products in their proper condition, is particularly essential. The extensive use of lighting in modern retail design can lead to significant heat build-up in summer (at a time when the passenger volumes are at their highest). Lighting is increasingly used in retail design to create the right ambience, to highlight particular areas of the shop and to increase the visual impact of product displays. With a combination of low-voltage and energy-efficient fittings, lighting design can reduce overall lux levels and provide varying contrasts. The use of intelligent lighting systems also allows the retailer to change or adapt the ambience at regular intervals over long or short periods of time.

Equipment and services

The fixtures and fittings used by airport retailers have many parallels with the high street. The nature of airport operations however places extra demands on retail equipment and services. The high volume of passenger traffic and the long opening hours of many airports (often extending to three shifts over twenty four hours every day of the year) can significantly reduce the operational life of retail equipment. Fixtures and fittings for use in airport retailing therefore need to be much more robust, requiring a much higher design specification. The airport authority in most instances will require the concessionaire to submit all equipment designs and specifications for approval and often demand a higher quality fit-out than would be needed in a domestic retail unit.

Because of the high quality of many of the products sold, airport retailers may require special equipment. For example, in order that the product remains in first-class condition, walk-in humidifiers may be required for the merchandising of Havana cigars. Similarly, to protect expensive cru class red wines, it may be necessary to provide a system of displaying a sample bottle in full light with the main stock lying in darkened bin units.

Owing to the high volume of stock depletion, shelving also tends to be wider than that used in a high street store. In an airport book shop, for example, twenty four inch, rather than twelve inch shelving is often used. Using the smaller, traditional shelving would increase the likelihood of stock-outs and lost sales. In the liquor and wine areas, as many as six shelf levels may be used with the lower two shelves being particularly deep to provide a high degree of back stocking. Shelves are frequently underlit to highlight the wide range of product shapes and colours, and to utilise the properties of the glass packaging to glow and sparkle.

Floor merchandising units such as counters and gondolas need to be mobile enough to allow retailers to change their layout in order to cater for special occasions or different passenger segments. For example, charter groups or travellers to the Far East may request that specific or additional products be stocked. Such demands require that units are sufficiently flexible and that the outlet's floor design specification includes a network of power and communication services to provide the necessary access for lighting and electronic cash register systems. Video viewing facilities, using modern touch screen technology and liquid crystal displays for product identification and branding purposes, can also be built into the floor gondolas.

The larger airport retail outlets may also be required by the airport operator to allow flight information screens to be placed in the store to allay customer fears of missing flights. Some airport stores have installed sound systems to provide in-store music and product advertising. While these systems can help to create a customer-friendly environment and generate advertising revenue, the operating authority frequently require that the airport's flight announcement and public address system can automatically over-ride any in-store system.

Airport authorities expect retail operators to use fully integrated electronic point of sale cash register systems that can provide a wide range of information, including sales turnover by product category, average spends by routes and by flight, and sales penetration rates in the various merchandise categories. The data provided would be used for concession auditing purposes and to assist the airport's commercial management in its efforts to maximise retailing revenues. EPOS equipment also needs the capability of handling a wide range of currencies and exchange rates, and the inputting of a valid airline flight number as the first requisite for initialising a sales transaction. This latter facility is a requirement of the Customs and Excise authorities who may also require the system to produce additional accounting and stock holding data for controlling the movement of bonded goods through the supply chain.

In the EU member states, an additional requirement is placed upon retailers to provide Customs and Excise with detailed sales transaction data in order to comply with the requirements of Vendor Control regulations. Under this provision, the vendor is responsible for ensuring that customers on journeys between EU member states can only purchase the legal allowance of duty/tax-free goods.

The large volumes of merchandise sold in airport retail outlets, particularly in the duty-free stores, make it essential that the retailer's EPOS system be linked to the warehouse, to the store and to the buying function. Because of the accuracy of such systems many airport authorities now specify the use of scanning equipment as a prerequisite to trade.

Despite the considerable level of security present in many airports, there remains a need for the retailer to provide a level of in-store security. Equally as air travel becomes affordable to a wider and more diverse passenger market, pilferage has become a growing problem for airport retailers. Video surveillance cameras, in-store security personnel, security tagging of merchandise and the use of alarmed detection gates have become features of airport retailing.

Merchandising of product

There remains considerable debate over the ideal location of the principal merchandise categories within a store. It is argued that purchases of tobacco and alcohol are more likely to be planned, while tax-free items such as perfume, jewellery and clothing are bought much more on impulse. The location of different merchandise categories will therefore partly depend upon whether duty-free or tax-free products are being sold.

Two schools of thought exist on the most effective method of merchandising these product categories. The first maintains that planned purchases should be located at the entrance to a store while impulse items should be located further inside. Consumers are then able to select their planned products immediately on entry, which will then allow them the opportunity to browse the store looking for impulse goods. Such an approach is predicated on the view that travellers will prioritise their merchandise selection and make their way directly to where it is located. Traditional duty-free products continue to have the widest appeal and account for the largest proportion of airport retail sales and profit. If positioned after the impulse items then customers are required to retrace their steps within the store.

A second school of thought takes the opposite view. Placing impulse items at the front of the store or close to the check-out means that consumers have the opportunity to browse without having to carry heavy or bulky products. While evidence of both types of layout may be found in airport retailing, the most commonly used approach places the duty-free consumable products at the first point of customer contact.

The location of merchandise within a store is not only dependent upon whether the product is classed as an impulse or planned purchase. Country of origin will often play a significant role. For example, in UK airports, Scotch whiskies will often occupy a prime location within a store, while in Ireland, home-produced spirits would have premium shelf space. Similarly, Lisbon airport allocates a premier location to port wines, while at Madrid and Barcelona airports Spanish wines warrant a significant presence.

Margin and unit cash value considerations will also play a central role in deciding the location of a product. Some would argue that the customer should be exposed to higher margin items and those that generate the greatest cash value early in the shopping activity. Others would place the most popular items to the front of the store and keep the specialist merchandise and low-volume sellers to the rear. Tobacco and alcohol products are generally placed towards the front of the store as they are high margin goods. Additionally, the high levels of customer brand loyalty associated with these products mean that customers requiring less decision time can quickly make their choice, and move on to view other merchandise.

Many airport duty-free retailers are now using computer software programs to facilitate merchandise space planning. Such technology is not new and has been used for a number of years by the large multiple food retailers. Programs such as 'Spaceman' produce detailed planograms for the layout of merchandise in store and provide illustrated shelf plans that specify the number of product facings as well as the positioning of each product type. The output from the software program is intended to assist retailers in maximising their return through the production of an efficient merchandise plan. This then optimises the volume of merchandise on display and minimises the replenishment sequence. As already noted, this is of particular importance in the airport environment where stock-outs can mean totally lost sales opportunities.

In order to utilise space management packages to the full, the retailer is required to input considerable product data. Such data will include full sales history, gross margins and cash price information for each product line, along with passenger traffic flow data and any unusual factors such as seasonality or special promotions. Because of the limitations on space at airports, retailers must balance the output of space planning programs (which may suggest high space elasticity for the top selling products and the allocation of multiple facings), with the need to present a wide selection of international brands. A number of internationally known products continue to remain the core merchandise group in the majority of duty-free stores.

Within the portfolio of global brands, however, there still remain strong regional and country wide preferences. The duty-free assortment is therefore tailored to fit the profile and nationality of the airport's passengers. The best selling brands on the passengers' home market are required to occupy a prominent position as well as be represented in the store. Many of the manufacturers of these brands have introduced line extensions especially for the duty-free market. De luxe and aged versions of spirits have been developed and presented in expensive porcelain or crystal containers. Other suppliers have developed brand versions which are exclusive to duty free and offer significant margin enhancement opportunities (see Chapter 4). There remain additional opportunities for merchandisers in the duty-free store to increase sales through product association, for example by placing cigars alongside the cognac section or wine accessories such as books, corkscrews and glasses in the wine cellar area.

The merchandising of perfume, skincare and cosmetic products replicates many of the techniques found within duty free. Internationally branded, prestige products with high market profiles comprise the merchandise assortment in most airport perfumeries. The fragrance houses have traditionally played a greater role in developing the merchandising strategy for airport retailers than have the liquor and tobacco manufacturers. The perfume manufacturers not only view the overall positioning of their merchandise as being of considerable importance, but also the location of their products relative to other specified fragrance companies. For example, Chanel may wish to be positioned between Lancôme and Clarins, while Estée Lauder might prefer a centre or pivotal position. Many of the larger fragrance manufacturers prefer to have a specific amount of space allocated to their brand and have their in-house designers work with the retailer to 'customise' their area. The retailer is under constant pressure from the fragrance manufacturers to stock new products and line extensions, but as with all other airport retail outlets, space is also at a premium in the fragrance store. The constant monitoring of turnover by product line is therefore essential to keep the merchandise presentation up to date and make sure each product provides an adequate return. As the space in airport perfumeries is finite, the retailer is often faced with eliminating a specific product line or perhaps even dropping a fragrance house completely.

The merchandising of non-consumable items in an airport mirrors the approach adopted in department stores. Airport tax-free stores are typically divided into broad merchandise sections such as jewellery, fashion, souvenirs, crystal and china, electronics and toys. The assortment profile of these shops would be similar to a full line variety store with a wide product range and little depth in any particular line. Because of the time constraints on passengers at airports, product range compatibility remains especially important to ensure that the merchandise makes a coherent and logical statement, which the customer can quickly and easily assimilate. Walters and White's (1987) assertion that the assortment profile must be planned to make a *coherent marketing statement* (p. 128) which is clear and concise and closely matches the customers' perceptions and expectations, seems particularly applicable to airport retailers.

In addition to stocking many international branded products, airport retail outlets carry generic products targeted at the traveller, such as travel irons, hair dryers, adapters for electrical appliances, batteries and films. These products are often bought by passengers on impulse and as such are frequently positioned close to the pay points. Airport retailers have found that having these impulse products merchandised at multiple points throughout the store maximises sales volumes.

As much of air travel is tourism driven, products that are typical of the country or city, for example, local food specialities or handicraft items, may warrant considerable merchandising space in airport shops. These souvenir style products can be difficult to merchandise as they tend to be very diverse in size and shape and are often not packaged. They are frequently merchandised in a dedicated area which can, for example, be designed to look like part of the local traditional market.

Jewellery and fashion items are usually merchandised in the same manner as in high street retail outlets. So too are electrical and electronic appliances, although the range would be focused on personal or small items, as most goods sold in airport outlets must fit into a passenger's hand baggage.

Merchandisers in airport stores must be careful not to present the time-pressed customers with too great a choice or price range and so make their purchase decision overly difficult. Equally they must ensure that products are clearly segmented by type and by price point to further minimise the possibility of customer confusion. Clear and well-designed product signage and graphics are also essential to allow passengers to make productive use of the time they have available to shop.

4 Retail Marketing within the Airport Environment

Introduction

Marketing is not new – despite the efforts of some to place marketing as a profoundly 20th century experience, others such as Brown (1995), have charted the development of marketing activity as far back as classical and pre-Hellenic Greece. The contribution (!) that academia has made to marketing thought remains more recent however. As Brown (1995) maintains, it was Drucker (1954) who first highlighted the pivotal importance of the consumer in the successful development of a business and who drew the distinction between the sales and the marketing function.

The late 1950s and the early 1960s saw the progressive development of the marketing concept as an academic subject. It was maintained that companies needed to move away from a production or sales dominated perspective towards a 'marketing orientation' (Keith, 1960; Levitt, 1960) where the changing wants of the consumer were identified and responded to. A focus on the needs of the customer through a marketing approach was therefore considered essential and could ultimately make the difference between business success and failure. As Brown (1995) notes;

> "The essential point is that marketing in general and the marketing concept in particular are widely regarded, by academics and practitioners alike, respectively to comprise the single most important management function and the key to success in business" (p. 33)

The Airport Customer

Airport authorities have not been slow to identify the importance of marketing as a means of gaining and maintaining loyal consumer support. The approach that many operators have taken has been to divide their customer base into two distinct groups and to market separately to each. This has led to airports focusing upon:

- the airline industry;
- travelling passengers and users of airport facilities.

In order to meet the needs of these two distinct groups the airport authorities have been heavily involved in marketing operations. There remains however some debate over which of these two categories represents the primary customer group. Some operators maintain that the principal focus should be upon the industry. As passengers remain the responsibility of the airlines they therefore represent a secondary customer segment (de Man, 1996). Because airlines have a degree of choice over which airport to use, operators should devote a significant proportion of the marketing effort to satisfying their needs and requirements. Creating a distinction between primary and secondary customers is in practice difficult to maintain, as the boundaries of responsibility between airline and operator are often obfuscated in the mind of the travelling passenger. The fact that the majority of consumers fail to distinguish between those services provided by the airport and those provided by the airlines has meant that operators have become responsible for 'over the tail' marketing. Attention has therefore gone beyond the immediacy of meeting the needs of the airlines to marketing to all who use the airport.

The focus of this chapter will be upon the strategies employed in meeting the needs of customers who use the retail and associated services within an airport. The strategies employed by an airport operator to encourage greater airline usage remain outside the remit of this chapter.

Numerous academic articles have commented upon contemporary consumers and the importance of responding to their changing needs. Increasing numbers of persons within Western society have become better educated, more widely travelled and demonstrate a greater willingness to experience tastes and values beyond their own, immediate culture. Airlines and tour operators continue to offer a plethora of options for the would-be traveller. Individually tailored activity holidays, Far Eastern travel and adventure breaks in remote and inaccessible locations all compete against the traditional charter holiday venues of Malaga, Benidorm and Ibiza.

With such an eclectic array of individuals now passing through European airports, retailers and airport operators alike have sought to provide customers with better choice, service and value. The extent to which airports have successfully achieved this strategy remains variable. Indeed some such as Gibson (1992) argue that many airports have failed to deliver on the majority of these criteria. Ballini (1993) argues that one reason for this is that airports have failed to understand the consequences of the transition from a regulated to a liberalised transport market. This change has created a series of specific marketing challenges for the airport authority. In particular, the movement away from a monopolistic to a deregulated environment has provided travellers with a much greater degree of individual choice over which airport to use. The consequent changes in consumer behaviour that this increased freedom has brought has prompted both airlines and airports to adopt a more proactive marketing approach (Table 4.1).

Table 4.1　*Consumer behaviour within monopolistic, regulated and liberalised markets.*

Monopolistic market	Regulated market	Liberalised market
The Consumer	*The Consumer*	*The Consumer*
Has no freedom of choice	Has limited freedom of choice	Can choose freely
Is affected by legal constraints	Can purchase services from a restricted number of suppliers	Can choose between a large number of different products and services
Can only purchase services from one provider	Can only partially influence his/her own loyalty level	Can choose between a number of different suppliers
Cannot influence his/her own level of loyalty	Provides easy-to-increase satisfaction levels because market provides customer with restricted options	Will provide loyalty only if satisfied
Shows loyalty independent of satisfaction		
Whether satisfied or dissatisfied, shows same loyalty levels		

Source:　adapted from Ballini (1993).

A liberalised air transport industry therefore places new demands upon the capabilities of the airport authorities. It becomes their responsibility to develop a clear market position that differentiates the airport from its competitors. In achieving this, it also remains their role to ensure that other parties involved in marketing the airport both understand the values of their consumer base and act in a way that is consistent in meeting their needs. In a retail context this means that an airport operator must

maintain a balance between active participation, constructive encour-
agement and the regulatory control of its concessionaires.

Achieving Differentiation

Individuals may be forgiven for assuming that there remains little to distinguish one airport from another. Each has an airside and landside segment, there are arrival and departure areas and if there are international
flights then there is likely to be a duty/tax-free facility. With the requirement to provide check-in facilities, car parking and car hire, one may
question the scope of an airport authority to provide a unique offer for
the travelling public. The objective for many operators, however, has
been to undertake a strategy that differentiates their airport from the
competition. Hooley and Saunders (1993) identify four ways in which
this may be possible.

Product differentiation

This strategy seeks to add value to the product or service provided to the
consumer. This improvement may be both tangible and intangible
depending upon the values of the customer group. Levitt (1986) maintains that a product or service can be viewed at four levels. The generic/
core product is the actual product itself, for example the liquor or tobacco
sold in duty free. The second level is the expected product, which refers
to what customers expect in addition to the core; for example, customers
in an airport may expect to pay by credit card or may expect sales staff to
speak English. The third level is the augmented product. This relates to
the additional services that are provided beyond what the customer may
expect; for example, gift wrapping of purchases or placing products in
protective packaging. The one issue with such initiatives is that they can
be copied relatively easily by competitors if seen to be successful. The
fourth level, the potential product, therefore remains important as it
requires the retailer to remain proactive by considering what other
facets can be offered in addition to those already being provided.

Promotional differentiation

This strategy attempts to use a wide variety of promotional methods in
order to create a strategy of differentiation. As will be illustrated later in
this chapter, airport authorities have, in line with many other commercial

organisations, reduced their reliance upon mass marketing techniques and moved towards a more focused promotional strategy.

Distribution differentiation

This strategy identifies alternative channels of distribution for a product or service. It may consist of having more extensive market coverage than the competition, using alternative formats to retail the product or having a different supply network. For example, the development of BAA's loyalty card not only allows the company to communicate with its customers but also provides them with the opportunity to directly promote new products and services.

Price differentiation

Hooley and Saunders (1993) maintain that a strategy of price differentiation can only be successful if the organisation enjoys a cost advantage or where there are barriers preventing organisations with lower cost structures competing at a lower price. Such a strategy has been successfully pursued by a number of airports in the Middle East. Abu Dhabi in particular has been successful in positioning itself as an airport that can offer a cost-based differential to consumers.

Marketing Strategy

One way of achieving differentiation is through the development of a co-ordinated marketing strategy. There exists an immense academic literature on the process of marketing planning and its contribution to the achievement of business objectives (Leighton, 1966; Kollatt *et al.*, 1972; Greenley, 1982, 1986; Brownlie, 1985; Ansoff, 1987; McDonald, 1994, 1995, 1996). While it is not the aim of this chapter to undertake a detailed review of the planning process, it remains useful to provide an overview of how marketing planning operates in practice. This will assist in a conceptualisaton of airports marketing strategies as well as provide an understanding of their marketing operations.

The starting point for the development of a marketing strategy is the Mission Statement (Figure 4.1). The Mission represents the guiding principles of the organisation and a statement of the nature of the company's business. Mission Statements vary from being succinct comments on the company's overall aim to more esoteric descriptions of the organisation's contribution to society. In the majority of instances, the scope of

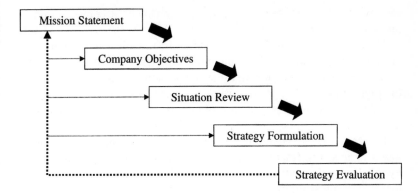

Figure 4.1 *The strategic marketing process.*

a Mission Statement is company wide and relates to the overall aims of the business. An example of a corporate level mission statement may include:

> "Aer Rianta wishes to establish itself as the best organisation in the world in the field of managing airports and associated commercial activities" (Annual Report and Accounts, 1991, p. 21)

Occasionally however they may focus upon a specific strategic area. A more targeted statement may be:

> "We will create a world class retailing experience for all our customers" (Retail Mission Statement, BAA, Annual Report and Accounts, 1993)

From the Mission develops the company's overall objectives. These represent tangible goals that the business has set itself to achieve within a particular timescale. Marketing objectives are then derived from these corporate goals. These marketing objectives are then translated into key result areas (such as market penetration and growth rate) which in turn may have a series of sub-objectives (e.g. product line extension, geographical expansion). The outcome of this process is a set of aims that have been derived from the strategic plan, that are achievable within the financial constraints placed upon the organisation, and compatible with the strengths and limitations of the organisation. The second stage identified by McDonald is the situation review, which is effectively an audit of the company. The aim is to identify whether any internal strengths or weaknesses exist that will assist or inhibit the marketing process; it requires a critical self-appraisal of the way in which the company under-takes its marketing planning.

The third stage in the process is strategy formulation and requires the organisation to develop and implement its marketing strategy. This typ-ically represents the element of the strategic planning process that carries the highest degree of risk, as it requires managers to make judgements and commit resources on the basis of imperfect information and market knowledge.

The final stage in the planning process requires the strategy to be eval-uated against the financial and non-financial criteria that have been established as indicative of the success of the organisation. Measures may include penetration rates, service and delivery levels or consumer sat-isfaction indices. BMS (1994) identified a number of factors that can be used within an airport, including customer satisfaction levels, customer loyalty, awareness and image, brand value, and the level of premium pricing.

The Exchange Relationship

A key determinant in the four-stage development of a successful market-ing strategy is an understanding of the concept of *exchange*. At its most simplistic, this refers to the exchange of values between two or more parties. Organisations manage to exist by negotiating exchanges within the marketplace (Anderson, 1982). In the business context, the provision of goods and services is continually exchanged for monetary reward and it is the role of marketing to manage these relationships. As Arndt (1983) maintains:

> "Hence the essence of marketing relates to the structuring and organ-ising of the exchanges between (and within) marketing institutions" (p. 44)

Such exchanges take a variety of forms, for example, they may operate on a short-term 'I win, you lose' basis; that is, one party disproportionately enjoys any gains made. A problem with this approach is that there exists little loyalty in the relationship, since consumers who have not enjoyed a good exchange relationship will seek alternatives. The highly com-petitive nature of business has meant that an increasing number of firms have looked to developing the longer term 'I win, you win' exchange relationship. The objective here is to increase the total value of the exchange by ensuring that both parties are mutually satisfied.

Achieving mutual satisfaction in the exchange relationship is reliant upon a four-stage cyclical relationship of understanding, creating, communicating and delivering values (Figure 4.2).

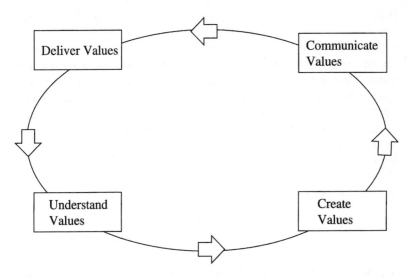

Figure 4.2   *The value cycle.*

As the values of consumers continually change, the need for a clear *understanding* of their attitudes, wants and desires remains a fundamental prerequisite for managing the exchange relationship. Recognising this change is necessary for the creation of new products, services and exchange environments. The *creation* of value can therefore be defined in its broadest sense to encompass all factors that influence a consumer's choice of product and store. In order to *communicate* the values that have been created for customers, a range of options remains available. One of the key developments in marketing communications has been the move away from a mass market to a more direct communications approach. While national and international TV advertising remains widespread, these have been supplemented by the use of direct mail and targeted media campaigns. The *delivery* of value refers to the actual exchange of goods or services between the two parties. It not only concerns the physical availability of product but also the less tangible, customer service levels that accompany the exchange. For example, the product knowledge and helpfulness of the sales staff may be considered an important determinant in delivering consumer values.

Understanding Value

If differentiation remains a key objective in an airport's marketing strategy, then the need for a detailed understanding of its consumer base is important for two reasons. First, it is important to ascertain what are the core values held by its customers. Secondly, there is a requirement to know what consumers expect in addition to the core product. Given that airports are a microcosm of cultural diversity, an environment within which one may expect to encounter almost every nationality, language and tradition, the need for this understanding would seem paramount. The challenge for both the retailer and the airport operator is to ensure that as consumer preferences or indeed consumer groups change, the retail offer within an airport is able to accommodate such developments.

Creating a *loyal* consumer depends upon having an understanding of the factors that influence purchasing behaviour. One could argue that because of the way in which an airport operates, retailers have a captive audience. Passengers are constrained as to where they can go and the activities they can undertake. For the retailer, this provides an opportunity to stimulate consumer awareness and encourage spending. The notion of a captive audience however represents an over-simplification of the trading conditions within an airport. While a retailer may be successful in the high street, this provides no guarantee of success within the airport.

For many, travelling by air continues to provide excitement as well as a degree of uncertainty over the correct procedures and operating schedules. Bingman (1996) identifies a number of natural stress points involved in using an airport. After arrival at an airport an individual has to go through ticketing, check-in, security and confirmation of departure. Each stage has the potential to place the traveller under a greater/lesser degree of pressure. Time often becomes an important factor in this process as the primary purpose of being in an airport for the overwhelming majority of persons is to catch an aircraft at an allotted time. Many passengers allow more than the recommended time in order to ensure that they make their flight and to avoid the congestion that has become characteristic of many airports. While the check-in times for some business routes have been reduced to less than half an hour, many non-scheduled and international flights have a two-hour check-in requirement. This, combined with the increased security measures present in airports worldwide, has ensured an increase in passenger 'dwell time', i.e. the time passengers spend within the terminal building prior to departure.

The uncertainty that surrounds many travellers when using an airport can have a negative impact upon spending. Some may suffer from 'gate lock', that is moving to the departure gate earlier than required and avoiding the retail offer on the central concourse. Some passengers will assemble at departure gates even when the aircraft has been delayed. Such behaviour illustrates the angst that many passengers feel when travelling and highlights the need for a detailed understanding of the consumer.

A multitude of different external, environmental factors therefore influences an individual's propensity to purchase duty/tax-free goods. One way of conceptualising these factors within an airport environment, is via the function:

$$PB = f(TE, L, PT, RE, PV)$$

where

PB = propensity to buy;

TE = tax environment (both direct and indirect taxation in the country of destination);

L = lifestyle (culture, social class, disposable income, leisure time available);

PT = product types (merchandise mix, range and depth, number of branded goods available);

PV = perceived value (the utility that accrues to the individual by purchasing or owning the product);

RE = retail environment (ambience of the airport, accessibility to retail outlets, store design and layout, staff attitudes and product knowledge).

The success of airport retailing is heavily weighted towards understanding these factors and this at least partly explains why airport authorities have established an enviable reputation for undertaking detailed survey work. Passengers' rationale for travelling, their purchasing intentions and their views on the overall ambience of the airport are among the most important factors that operators seek to identify. For example, BAA interview over half a million passengers every year across its seven airports, while Schiphol airport undertakes 100,000 quantitative interviews and 600 qualitative interviews with passengers. In the majority of instances the airport authorities subcontract this exercise out to a third party who will be responsible for the collection of data as well as a basic analysis of the results. A number of airport authorities also require the subcontractor to supply all the data to them after the basic analysis so they may undertake a more detailed statistical dissemination.

The market research data collected on consumers in airports falls into three broad categories. First, traffic route information identifies a passenger's nationality and country of residence, frequency and class of travel, length of stay and reason for travelling. This information can be cross-tabulated against various demographic data and quota sampled by airline. In addition to providing information relevant for retail use, such data may be used to market an airport to the airline companies and assist in the development of new routes.

The International Air Transport Association (IATA) survey monitors basic standards in 25 different airports and provides a benchmark against which operators can be judged. A number of airport authorities use similar consumer satisfaction surveys as a means of evaluating their service and quality standards relative to those identified by IATA. Passenger attitudes are gauged on a wide range of different criteria including check-in and arrival facilities, baggage handling, and restaurant and catering provision. Aer Rianta, for example, interviews between 800 and 1000 persons per month on their attitudes to the service provided at Dublin airport.

The third area of research is retail specific and questions consumers on their purchasing habits and behaviour. It attempts to identify the type and quantity of product bought and disaggregate this by demographic group. The number of consumers purchasing product relative to the total number travelling, i.e. the penetration rate, is used by airports as a measure of the success of their marketing campaign and ultimately their retail activities. One common objective of this type of research is to discover why consumers do not purchase while at an airport and to identify the factors that would encourage them to do so. Typically the results of these surveys will aid the airport in its future commercial planning activities and will influence areas of decision making such as unit location and retail tenant mix. In addition the findings from such surveys may be made available to the concessionaires in order to assist in their marketing activities. These are often disseminated on a quarterly or six-monthly basis through a presentation and written report by the airport operator.

Airport authorities tend to have strict regulations governing who is permitted to collect passenger data. Apart from their own market research the number of additional surveys tend to be limited to either large-scale international organisations, e.g. IATA, or confined to specific industry issues, eg. the abolition of duty free. This restriction also extends to data gathering by concessionaires. At Heathrow and Schiphol, for example, retailers are prohibited from conducting their own market research within the terminal without prior permission of the airport operator. Such permission is only forthcoming if the retailer is able to demonstrate a clear need for the additional data. Other than in these circumstances, the concessionaire is reliant upon the airport authority to provide the relevant market research information.

Technological advances in data gathering techniques have provided both the airlines and airports with the opportunity to adopt new measures for understanding customer behaviour. Through their frequent flyer programmes and the data made available through their ticketing and central reservation systems, many airlines have been successful in securing database information that can be used for marketing initiatives.

Airports have almost no access to these data sources and even those airports currently involved in providing ground handling and facilitation services at airports gather little additional airline passenger data. Such data are jealously guarded by the airlines for competitive reasons as much as for their commercial value. However, through the widespread use of modern Electronic Point of Sale (EPOS) systems, airports now have the opportunity to access a significant data source. Almost all airport retail operators are required by their Customs and Excise authorities to record a passenger's airline flight number on the cash register. This has allowed operators to link sales data to specific flights and to match such data with the associated route profiles (e.g. passenger nationality, purpose of journey, class of flight etc.). Many of these EPOS systems also allow for the till operator to ask the customer a number of simple closed questions with a limited number of reply options. Responses can be linked to special keys on the cash register and will only require a simple keystroke by the operator to input the data. Despite this greater availability of data, many airport authorities still do not have access to all information, owing to the clear delineation of functions imposed by the retail concessionaire structure. Only those airports that operate their own retail outlets or those with management contract arrangements can guarantee complete access to all EPOS information.

A small number of airport authorities have attempted to extend and widen their data gathering techniques by developing a loyalty card similar to those operated by the airlines. This rewards individuals on the basis of the amount that they spend at the airport retail outlets. For example, BAA operate a loyalty card scheme that can be used across all seven of its airports, while SAS Trading have a tax-free loyalty programme that operates in all its retail outlets. Individuals receive points with every purchase in the airport, and these points can then be used for price reductions or redeemed for shopping vouchers, exclusive merchandise or air miles.

In the late 1980s Dublin airport started a wine club for their duty-free shoppers which offered special wine selections for members at discounted prices. Activities included regular news and information publications and a yearly wine tasting fair at the airport. This provided the airport with a database of regular wine buying travellers which was subsequently widened to include all those passengers using the airport business lounges, pre-purchased car parking and those pre-ordering duty/tax-free goods from the airport shops. While loyalty programmes are not new and have operated in other areas of the retail sector for some time, their widespread use within airport retailing is uncommon. The benefits of loyalty cards are numerous. Not only do they provide an effective means of understanding consumer purchasing behaviour, they permit the development of a direct mail database and provide a much more focused and cost-efficient means of communication.

One key outcome that has stemmed from a greater understanding of consumer needs and aspirations has been the move away from the traditional mass market approach to the development of a more targeted marketing strategy. Kotler (1991) maintains that a successful segmentation strategy requires four criteria to be met. First, the segments have to be measurable, second, they must be able to be reached cost effectively, and third they need to be substantial enough to be worthwhile targeting. Finally they need to be actionable, that is the company is able to formulate specific plans to serve the segment.

A segmentation approach provides the retailer with the opportunity to formulate a defined positioning strategy that will differentiate it from the competition. While some have noted the dangers of over-segmenting the airport industry (Humphries, 1996), such a strategy provides a number of benefits. In addition to improved customer service and better communications, it provides a more tightly focused merchandise range that eliminates inappropriate product groups and allows the better utilisation of in-store space.

Airport authorities have sought to segment their offer and to match the retail provision more closely to consumer needs. Undertaking a segmentation strategy is not new and has been identified in a wide variety of different sectors such as footwear and clothing (Walters, 1988), grocery (Sparks, 1993) and retail financial services (Harrison, 1995). The consumer market may be segmented using a variety of criteria. The three most common approaches used are demographic (age, gender, ethnicity), socio-economic (income, family size, employment) and psychographic (lifestyle, behaviour, personality). The airport industry in the main has pursued a strategy of demographic and socio-economic consumer segmentation. Four broad categories of airport customer have been identified as being most applicable to an airport environment, as detailed below.

Passengers

Passengers obviously represent the most important consumer segment for retailers. The extent to which this category may be further segmented has grown with the increased volumes of passenger traffic. Because of the customs and security requirement to record all passengers travelling, the information that exists on this segment is both more available and more accurate than one would expect in other forms of retailing. The broad sub-segments that passengers have been further classified into include:

Domestic v. international v. transit: This represents one of the most important segmentation categories from a retailer's perspective. Airports with a high proportion of international travellers benefit from the fact that passengers tend to spend longer in the airport, have the opportunity to purchase duty-free products and have a higher likelihood of being accompanied when travelling to and from the airport. The wide range of retail facilities at Schiphol has been attributed to the fact that 99.5% of its passengers are travelling internationally. In contrast, the limited retail operation at airports such as Vigo, Bilbao, Stockholm and Oslo has been attributed to the low number of international passengers (Doganis *et al.*, 1994; Gray, 1994).

Intra-EU v. non-intra-EU: Profiling the passengers travelling between EU member states has been useful to retailers, primarily for marketing purposes. However, with the threat to abolish duty/tax-free sales on intra-EU flights in 1999, the identification of intra-EU passengers has taken on a new significance. Retailers and airport operators have calculated the potential loss of revenue stemming from its removal. In 1995, intra-EU duty/tax-free sales at airports amounted to $1.8 billion, and by the proposed abolition date of 30 June 1999 lost sales would be expected to be in the region of $2.6 billion. Airports currently account for over 52% of total intra-EU duty/tax-free sales, with ferries and airlines accounting for the remainder (ETRF, 1996).

Short-haul v. long-haul; scheduled v. non-scheduled: The types of product bought and the amount spent within an airport will depend upon the type of journey passengers are embarking upon. For example, long-haul passengers spend proportionally more than those on short haul, and moreover they tend to buy gift items, jewellery and fashion in preference to tobacco or liquor. Persons on scheduled flights spend less time within an airport than those on non-scheduled flights. This is primarily due to a combination of shorter check-in times, a greater understanding of airport procedures and a greater familiarity of the terminal layout. A combination of such factors allows a more accurate calculation of the time needed to catch a flight.

Scheduled passengers have a higher propensity to purchase duty/tax-free branded products than charter passengers. At Birmingham airport in the UK, for example, research on the average spends of different types of passenger indicated that those on a charter flight had a lower propensity to purchase liquor but a higher likelihood of buying tobacco and perfume products. These different purchasing patterns are partly influenced by the composition of passengers travelling as well as the destination of the charter flight. Many charter locations are in Southern Europe where duties and taxes on liquor are traditionally low. The strong product loyalty exhibited by tobacco consumers and the fear of being unable to find their favourite brand at their intended destination may also influence purchasing patterns.

Business v. pleasure: Business passengers represent a wealthier cohort than the average population. This is primarily due to the disproportionate number of professional and managerial persons that comprise the category. These travellers have higher levels of disposable income and identifiable spending patterns. Furthermore, business passengers are considered to have only a limited time to shop in the country in which they are visiting. Airports may therefore represent one of the primary opportunities for them to purchase products. However, as a segment they represent a challenge for the retailer. Business travellers are less likely to browse, may regard shopping as unnecessary and often spend less on any single journey owing to their high frequency of travel.

Nationality: In addition to dividing passengers on the basis of their class of travel and their motives for travelling, individuals may also be segmented according to their nationality; for example, in Europe,

Greek and Norwegian visitors have the highest per capita per day spending. This often reflects a price differential for branded products in the home and visited countries. Humphries (1996), for example, notes that Japanese travellers spend up to ten times more than other nationalities, with an average outlay of US$200 on duty free per trip.

Purchasing patterns are not static and there is evidence to suggest that, because of the continuing recession in Japan and the resultant fall in the value of the yen, the importance of Japanese customers will diminish relative to other nationalities. For example, Korean travellers have been identified as representing an increasingly important consumer market. Such change will impact upon product demand as different nationalities display preferences for different merchandise categories. For example, Scandinavian visitors have a greater propensity to buy spirits and fine foods; French passengers buy fashion knitwear and quality malt whiskies; while the Japanese purchase general cosmetics and cognacs.

Access to such detailed passenger-related information allows the airport to develop a detailed customer profile. For example, in 1993/94 Gatwick airport had 20 million passengers who spent an average of 1–2 hours in the airport. The passenger composition was: 48% aged 25–44, 57% male, 81% ABC1, 72% UK resident, 11% American, 19.2% on business and 38% had travelled 1–3 times in the previous 12 months (Weber, 1995)

Airport authorities, however, have not focused purely upon passengers but have developed their tenant mix to focus upon other market groups. Prompted by the threat of duty-free removal in 1999 and a realisation that non-travellers represented an untapped potential source of revenue, airport operators have sought to widen the appeal of their retail offer beyond that of departing, arriving and transit passengers.

Airport Staff and Airline Crews

Although not extensively developed, convenience-based retail developments have sought to encourage staff and airline crews to use the airport as the primary destination for shopping. Because of the location of many airports away from the city centre, access to shopping facilities remains limited. In order to fulfil the needs of those working at the airport, operators have developed tenant mixes that include retail services such as supermarkets, laundry, photo-processing, pharmacy services and shoe stores.

Meeters and Greeters and Taxi Drivers

Meeters and greeters and taxi drivers may regularly use the airport. They have therefore been identified as potential consumer segments, and landside facilities such as bookshops and catering facilities have been improved in order to appeal to this market. Estimating the actual number of individuals meeting persons at the airport remains difficult and will relate to the profile of the airport. For example, Irish airports have a significant number of passengers who will be visiting friends and relatives (VFR). There exists a greater propensity for entire families and groups of friends to meet and greet their relatives at the airport. Similarly, at Birmingham airport the large Asian community in the region leads to a high level of VFR traffic and meeters and greeters at the airport. While such examples exist they do remain atypical of the majority of airports; Shaw (1993) maintains that many airports work on the principle of a 3:1 ratio. An airport of 12 million passengers would therefore expect an additional 4 million meeters and greeters.

Local Residents

The importance of an airport as a primary leisure and recreational destination should also not be underestimated. Humphries (1996), for example, notes that Gatwick is already among the top five fee-paying attractions in the south-east of England with over a half a million visitors to the spectators' gallery a year. Proposals are underway to capitalise on this popularity with the development of a theme park next to the airport.

Schiphol Plaza in The Netherlands is a landside shopping complex situated in the main terminal building and through which all arriving and departing passengers must pass. It comprises 35 retail outlets and has 5400 sq. metres of floorspace. Its development has been supplemented by an integrated transport network that services both the airport and the shopping facility. This, combined with extended opening hours and in particular Sunday opening, has meant that the airport has been successful in targeting and attracting the local population.

The segmentation strategy implemented within airports has been one based upon demographic and socio-economic criteria. There have been few proposals to develop a more sophisticated method of consumer profiling. Ballini (1993) perhaps remains an exception to this in his attempt to develop a behavioural profile of airport consumers. Such an approach in itself is not new, Stephenson and Willett (1969), for example, identified four separate shopping orientations while Cathelat (1990) attempted to develop a classification system based upon consumer lifestyles. McGoldrick (1990) maintains that the value of such an approach lies in its ability to relate directly to retail patronage activity.

Ballini (1993) segments airport consumers on the basis of the loyalty exhibited towards the airport. The extent to which such behaviour manifests itself will be a function of the airport's ability to deliver value through the exchange process and the degree of market liberalisation. The main categories of consumer behaviour are:

- *Loyalists*: these represent the most important group of consumers within an airport. Based on their past experiences, they continue to choose the airport as a departure or arrival destination.
- *Defectors*: these include persons who may be satisfied with the current provision of an individual airport but are not so satisfied to the extent that they will automatically choose it.
- *Mercenaries*: these consumers tend to be price or fashion conscious. They exhibit no loyalty towards a particular airport.
- *Hostages*: these consumers have no choice over the airport they use and are compelled to accept the services and facilities that are offered.

The extent to which passengers exhibit the above traits will in part be dependent upon the choice of airports available from which to travel. For example, passengers living in Copenhagen will have access to intercontinental flights from what is effectively their local airport. Because of this availability, the incentive to hub through other European airports will be considerably lessened. In contrast, passengers using regional airports may have a much greater degree of discretion over which hub to use. For example, intercontinental passengers flying out of Central Scotland will have a choice of using either Edinburgh or Glasgow airports and transiting through Schiphol, Heathrow, Copenhagen or Frankfurt. Airport authorities therefore cannot rely upon their size to be an indication of the loyalty of their consumer group. This not only highlights the importance of consumer research methods for understanding attitudes, habits and opinions but underlines the need for airport operators to continually create new and attractive propositions for the customer.

Creating Value

The adage 'what is new in terms of customer service today will be consumer expectation tomorrow' is often used by marketing academics.

The point that derives from this is that airport operators and concession-aires alike have to be proactive in developing their existing offer as well as looking for new means by which value can be added to the retail pro-position. While creating value is often associated with new product devel-opment (NPD), focusing solely upon NPD excludes functions that are central to the creation process. A more holistic definition would include the design and layout of the terminal, the composition of the retail tenant mix and the correct use of in-store atmospherics. This broader, more encompassing approach avoids the narrow delineation of functions and provides a more accurate understanding of value-creating activities. The importance of retail design and planning and tenant mix control within an airport terminal was the focus of Chapter 3; its contribution to value creation highlights how a successful marketing strategy requires a cross-functional approach.

Central to the creation of customer-related value is the product offer. While the majority of products sold within an airport can be classified as 'dynamic' (i.e. their shelf life is longer than the replenishment cycle) (Harris and Walters, 1992), because of the diverse range of customer segments within an airport, it remains impossible to identify a single or generic product strategy. The approach taken to the development of a product strategy will therefore vary by company. There however remain two key features that influence the merchandise decisions of an airport retailer. First, because the majority of airport retail outlets in Europe are situated airside with most of the products being sold duty and/or tax free, consumers expect to purchase goods at lower prices than downtown. Secondly, consumers expect to have access to a wide assortment of inter-nationally recognised products. For example, in Northern Europe the overwhelming demand is for branded alcohol and tobacco products, while in southern European airports, tobacco and perfume products would be more dominant.

Within the alcohol and tobacco market the number of new products introduced every year is limited. Consumers tend to display regular buy-ing patterns and undertake little impulse purchasing. New product development therefore remains relatively low and the creation of new brands is supplier rather than retailer driven. In the tax-free fragrance market, internationally branded products are also in greatest demand. The level of impulse purchases is, however, greater than in duty free and new product development is consequently higher. While retailers are required to stock the newest brands, there is again little evidence of a proactive approach on the part of the retailer. The introduction and adop-tion of new products tend to be almost wholly supply-side phenomena.

The growth of organisations such as Nuance/Allders, DFS/LVMH, World Duty Free and Aer Rianta may stimulate increasing levels of co-operation between retailer and supplier in the development of new products and line extensions. Because of the volumes of products that are now sold through duty and tax free, it becomes viable for manufacturers to create product ranges exclusively for an individual retailer. The recent launch of cognac products such as 'Fleur' targeted specifically at female consumers in the Far East is the outcome of such collaboration.

Consumer preferences dictate that retailers stock the latest brands; this not only exposes the product to an international market, but also allows suppliers to monitor demand against specific market segments. Not all retailers pursue a strategy of stocking the latest product releases. For some retailers of high-value items, such an approach would be inappropriate. With the sale of tax-free watches, for example, the objective is not necessarily to provide the consumer with the most up-to-date make and model but to provide a product that has been established on the market for a number of months, possibly even years. This is primarily because individuals buying high-value products undertake comparison shopping. They will either visit the tax-free outlet on more than one occasion before buying, or will have compared a number of landside prices previously. The retailer must establish in the customer's mind that the offer is genuine, dependable and that post-purchase contact and servicing will be available.

Unless travellers are frequent flyers they will have little time to acquaint themselves with the layout and design of the store. Consequently the merchandise mix within an airport shop needs to be more restricted with a limited *choice* of products on sale and a high speed of customer service. For example, the stocking policy of an airport bookshop will differ significantly from its high street counterparts. In addition to having a higher proportion of gifts and souvenirs, there will be fewer book titles available in the airport with a focus upon key subject areas. The objective is to make a strong product statement through the stocking of key authors. For example, while a domestic store may stock the top twelve crime authors, an airport might only stock the top four. However, this reduced range is partly compensated by ensuring that all the titles of these top authors are stocked. The focus upon depth rather than breadth in an airport is further illustrated by the quantities ordered. Whereas a high street retailer may hold 150 copies of a new best-selling publication, an airport may hold up to 4000 copies of the same title.

In line with the stated aim of many airport authorities, retailers have sought to develop a product strategy that complements the market position of the airport itself. *Product quality* has become a significant factor in the merchandise assortment and is reflected in the wide variety of international brands now available within an airport. Retailers seek to further reinforce this image through the provision of quality fixtures and fittings. For example, chocolates are often displayed in air chilled cabinets, which are furnished to a high standard in natural timber, copper and brass. However, given the high rate of stock turnover and the fact that the store is air conditioned, refrigeration is not required to maintain product quality. Nevertheless, it is provided by the supplier as part of its promotional support and used by the retailer to reinforce the perception of quality and perfect condition.

The high demand for international brands among airport customers would suggest that new products, developed specifically for the duty/tax-free market, would be much more difficult to establish. Retailers and suppliers therefore face a dilemma, since the investment required to build a brand solely for duty or tax free is significant both in terms of time and cost. At the same time there is a need to differentiate the product offer from the competition and provide an *exclusive* retail offer. This balance has been achieved by 'piggybacking' new products on brands that are already available in the main domestic markets. Such a strategy requires substantially fewer additional resources as the consumer is already aware of the brand name.

For example, the Pernod Ricard duty-free subsidiary World Brands developed a line extension to its Jameson Irish Whiskey range with the introduction of Jameson Gold. While consumers were familiar with the Jameson name, the product was exclusive to duty free. Similarly, Revlon segmented the cosmetics market and introduced 25 different duty-free multi-packs aimed at different groups of international traveller. Utilising existing brand names to build exclusivity remains an important marketing strategy as time-pressured travellers are able to immediately identify the core values of the product. Philip Morris, for example, launched Battistoni cigarettes as an exclusive product for duty free, with no equivalent product in the domestic market. The lack of brand recognition was one factor that led to it being withdrawn from sale.

Design has become the fourth consideration within an airport retailer's product strategy. In addition to having an aesthetic value, it has been necessary for designers to develop products that are portable enough to be carried, light enough to meet weight restrictions, compact enough to be classed as hand luggage and durable enough to accommodate the demands of air transport. For example, manufacturers have begun to supply spirits in plastic PET litre and half litre bottles. The objective has been to keep the appearance and durability of a glass bottle while at the same time making the product lighter to carry. Copenhagen airport has overcome the difficulty of selling bulky bed linen by packaging such products under a power press. Duvets, for example, are compacted into boxes that make them portable enough to carry on aircraft as hand luggage and light enough to fit in overhead lockers.

Service is the fifth factor within a retailer's product strategy. Retailers in creating value are required to decide on how their products are to be sold. Operating a self-service style of operation provides time-conscious consumers with the freedom to browse the store at their leisure. It also allows the frequent traveller who is familiar with the shop and short of time to rapidly select and pay for goods. Some element of counter service is necessary for customers unfamiliar with the range and complexity of some product categories. For example, cosmetics and skincare suppliers often provide assistance for male customers who are making purchases for their partners.

Time continues to be a critical factor in influencing an individual's propensity to purchase and has led some retailers to operate different sales methods in the airport compared with the high street. For example, one company that manufactures hand-made chocolate normally presents its products on trays and allows consumers to indicate which variety they prefer. However, in order to minimise the time involved in selecting and packing, the products on sale in the airport are sold on a pre-packed, self-service basis. To facilitate customer choice, up to eight different package weights are offered with the selection and flavours clearly illustrated on the outside of the pack.

Pricing remains central to a retailer's merchandising strategy. One perception that the operators have sought to overcome is that airport shopping is necessarily expensive (Lundqvist, 1994). Airport pricing varies both by country as well as by airport and will be influenced by the margin negotiated with suppliers, the prices of products elsewhere and, in cases where there is a high reliance upon a particular route, the prices at the destination airport. Landside shopping in German airports, for example, is considered to be expensive by many German consumers. This is primarily due to the near-monopoly position airports have held over week-end shopping in Germany when retailers have traditionally been forced to close on Saturday afternoons and Sundays. A 1995 survey conducted by Mori for BAA illustrated significant price differentials between major product categories within European airports, and moreover many consumers did not trust the pricing policies of airport retailers (BAA, 1995; Weber, 1995).

Pricing can however be used as a major source of differentiation. Products sold in airport shopping malls in the USA, for example, are typically 10–15% more expensive than downtown. Pittsburgh International airport has targeted domestic travellers and provided a guarantee to charge 'regular mall' prices. Since implementing this strategy, sales within the terminal have increased by 250%. In the UK, BAA sought to change consumers' attitudes to airport retailing by providing them with a Value Guarantee. Costing approximately £1 million per annum, the objective of this strategy was to communicate to customers that duty-free prices would on average be 50% less than high street prices.

Space within a trading unit is often limited and it is the responsibility of the retailer to maximise the square foot return. Product range reviews are therefore a necessary part of a retailer's product strategy. These are carried out every 3–6 months, often with a major review every twelve months. The objective is to identify those items that no longer provide a significant enough profit or sales contribution and can be removed from sale. These are then delisted in favour of new products. The criterion for delisting a product varies. In the majority of cases it is based upon the previous levels of sales, however the retailer may be compelled to remove a brand if the supplier no longer wishes to support it in terms of advertising and promotion.

In the perfume market especially, there is an increasing trend by manufacturers to introduce new fragrances almost every year. Significant pressure is placed upon the airport retailer to stock these new lines both from the manufacturer and from passengers. The limitations on space provide retailers with negotiating leverage as they are able to place a premium on making space available for new lines and thus gain more advantageous terms from the supplier.

Whatever product policy is pursued, it remains accurate to suggest that retailers have a considerable degree of autonomy over their stocking policy. Whereas the airport authority may limit the amount of market research that can be conducted, or place constraints upon the level of in-terminal communication (see below), normally there is little attempt to influence the merchandise mix of the concessionaire. Some exceptions to this rule do exist however. For example, it would be the norm that the concession agreement specifies that leading international brands (top ten or twenty) be stocked. Products such as Marlboro cigarettes and Chanel perfume could not be withdrawn from sale without either a valid business reason or the specific agreement of the airport operator.

Communicating Value

The primary purpose of the majority of people using an airport is to travel to another destination rather than to shop. Many travellers will be in a lesser/greater state of anxiety so the first priority of the airport's communication strategy will be to overcome the pressures that consumers find themselves under when using an airport (Bingman, 1996). Communications therefore operate at a number of levels. In the first instance, airport operators must ensure that embarking passengers are adequately informed about all aspects of their flight. Directional signage must be clear, unambiguous and easily understood, regardless of an individual's language or nationality. Equally, all aspects of the flight (departure/arrival times, baggage retrieval, delays, changes in gates etc.) must be clearly displayed.

As their next priority, airports will want to provide information on the range of passenger services and commercial activities available. In particular, the retail offer within the terminal will merit considerable emphasis. The airport authority may communicate this within the terminal, within the airport environs or prior to arrival. Retailers and suppliers are also proactive in their advertising strategy and seek to communicate with customers in-store, in-terminal or outside the airport.

If handled appropriately, a communications strategy has the potential to reinforce the values of the airport, highlight its principal features of differentiation as well as provide a significant financial return by stimulating additional sales and generating advertising revenue. Advertising in the form of illuminated wall signs, showcases and product displays, situated both inside the terminal buildings and outside in the general environs of the airport, can all generate considerable income. BAA, for example, generated approximately £15 million in advertising revenue in its seven airports (Gray, 1994). Such activities are usually operated by specialist concessions such as More O'Ferrall or Sky Sites. In return for a percentage of all revenue earned, the company will take responsibility for the management of all advertising media.

The starting point for understanding the commercial communication process within an airport is the interface between the retailer and the airport authority. This may operate at both a formal and informal level, with one of the most common methods being scheduled meetings. At Schiphol airport, for example, all the airside concessionaires meet together with representatives from the airport authorities on the first Wednesday of every month. A member of the airport authority always chairs the meeting and the group discusses both strategic and operational issues relating to retailing at the airport. In addition, the airport has established a series of working groups to focus upon specific retail issues. Again the airport operator chairs these meetings with a number of concessionaires represented on each. One role of the working group is to develop a promotional strategy for the airport, which centres upon planning special offers, ensuring co-ordination for special events such as Christmas and advertising the shopping facilities as a whole. Concessionaires contribute financially to all promotions (0.5% of the previous year's turnover), the budgeting of these activities being the remit of a second working group. All report to the main group meeting.

Where an airport has both a landside and an airside facility then there is often the need to communicate these two elements separately. Again Schiphol airport represents a good example of such a strategy. The target market of the landside shopping centre is primarily local during the weekend and airport staff during the week. The Plaza has its own logo and is managed by a different working group from the airside retail.

The airport operator has a range of established options upon which to draw in order to promote the retail facilities available at the airport. Crosier (1994) notes that these can include advertising, publicity, packaging, direct marketing, sponsorship, personal selling and sales promotion. Some methods, such as advertising, are more suitable for creating a general awareness while personal selling may be necessary for the long-term adoption of high-value products. The choice of promotional mix used will also depend upon the target audience, the message that needs to be conveyed, the costs involved, the need for effective measurement and the level of control that can be exercised over the outcome (Crosier, 1994).

As noted above, the opportunity for retailers and airport operators to communicate with potential customers occurs at three stages: outside the airport; within the terminal building; and within the store itself. The challenge for the airport operator is to utilise the communication media highlighted by Crosier (1994) across these three environments in a co-ordinated, systematic manner in order to deliver a clear and consistent message.

Communication Outside the Airport

The operator has two objectives in this context. First, to communicate to the airline industry the potential benefits that would accrue from choosing its particular airport. The competitive nature of the air industry means that airport authorities have to take a much more proactive role in attracting the airlines and the travelling public. Many airport operators maintain that their primary customers remain the airlines (de Man, 1996). To communicate the benefits that accrue from using a particular airport, operators have undertaken an aggressive direct marketing approach towards airlines and have also sought to involve potential destination airports in joint studies of route possibilities and co-operative promotional efforts. These have focused upon emphasising criteria that are considered to be important for the airline's own delivery of value. Factors such as the number of airlines handled, the ability to obtain traffic slots, the ease and speed of transfer, quality and cost of ground handling, and the level of marketing and promotion are among the criteria considered important when choosing an airport.

Many airports target tour operators because of their close approximation to the traveller. Typically an airport contracts the tour operator to put together a charter package with the airport providing financial support at so much per passenger. These support arrangements can alternately be based on the provision of a shopping voucher to a specific value, offering a particular item at a reduced price or possibly offering a percentage discount on particular lines of merchandise. Vouchers and special promotions would only be valid in the airport's duty/tax-free shops. From the tour operator's perspective, such incentives can provide a competitive advantage by making its charter package more attractive to the consumer. From the airport's perspective, the scheme gives passengers the added incentive to enter the airport shops with the possibility of extra impulse purchases.

In the case of travel agents, an arrangement is often made to insert a price list or shopping catalogue into the passenger's travel pack or have the airport shop advertised on the back of the ticket wallet. In order to gain the loyalty of travel agents, some airports have gone a stage further, Schiphol, for example, has operated a 'diamond roadshow' in 50 European cities. Through the use of videos and presentations, the objective has been to reinforce in the minds of the travel trade the potential benefits of using Schiphol as a departure/arrival/transit airport. Such an approach provides direct access to influential 'gatekeepers' in the air travel industry, while at the same time remaining a cost-effective marketing approach for both the aviation and retail side of the airport.

By marketing to travel agents and tour operators, the airport is able to communicate indirectly with passengers. The aim is to highlight the variety of goods and services available in the airport and allow passengers to make their choice of merchandise in advance of their arrival. Given the increasingly competitive nature of the duty/tax-free market, such a strategy has grown in importance. Some companies, such as Premiair airlines in Scandinavia, have also begun to target the consumer before they have arrived at the airport. On receipt of their flight tickets, individuals are provided with a catalogue from which they are able to choose duty/tax-free goods. Ordering may be done over the telephone and the products are placed on the passenger's designated seat in the aircraft on the day of departure. A key to the success of this strategy has been the ability of the airline to initiate the communication process at the pre-arrival stage. Through the use of information technology the airline has succeeded in providing a seamless delivery service.

The original approach of many airports was to externally advertise the airport through mass market media, typically in broadsheets and magazines. As a marketing medium the advantage of this form of advertising was the widescale coverage that it could provide. Crosier (1994), for example, notes in Britain alone there are 35 national newspapers, 110 regional daily newspapers, 1250 regional weekly newspapers, 900 free-distribution local newspapers, 1500 consumer magazines and 2500 specialist magazines. While allowing the airport to create a general awareness of its retail provision, the main disadvantages with general advertising have been the cost and the inability to target particular market segments. A more focused approach has therefore been attempted by a number of airports and a greater variety of media have been used to communicate the products and services on offer. These include;

Catalogues: These have become an increasingly important method of communication with travellers. Some operators print detailed tax-free catalogues that are then exported to foreign agents or distributed by direct mail to holders of loyalty cards or members of frequent flyer clubs. For example, the duty-free catalogue published by Schiphol airport is distributed worldwide through the KLM network of offices. In addition, the airport produces compact product catalogues for distribution through travel agents. In this way, information on the full range of the airport's retail offer is available to travellers prior to their arrival.

Despite airports attracting a disproportionate number of business travellers with a higher than average level of disposable income, their familiarity with product availability and their frequency of travel make them a highly discerning market segment. In Ireland, Aer Rianta have sought to overcome these difficulties by targeting the resident business traveller with a direct mail catalogue. Using information from a range of sources, including its lounge membership and wine club database, it has targeted business travellers with a specialist brochure, which is mailed directly to the individual. Prior to any international travel the passenger is able to phone in an order which will be ready for collection on departure. The company considers this method of distribution differentiation to have been a success, with average sales double those of non-targeted business travellers. Such a system has the added advantage of allowing other members of the target passengers' families to see what is available at the airport and add their requirements to the purchase order.

Internet: The rapid growth of the World Wide Web has provided a number of airports and airlines with a new method of communication. Easily accessible for anyone with a personal computer and modem, the Internet has been used by airports for information provision. Web-sites may include basic information on the structure and operation of the airport, recent press releases, annual reports and flight information. This medium would also allow a regularly updated listing of what merchandise and special offers are available at the airport shops.

In-flight magazines: Interestingly, this medium does not represent a major communication channel for the airports. Primarily this is due to the large number of airlines that use an airport. To advertise regularly in each in-flight magazine would be prohibitively expensive. Operators therefore tend to target only those airlines that provide the largest levels of passenger flow. For the airport operator these magazines offer a long-term image building medium rather than a proactive source of short-term business. Shannon airport in Ireland uses in-flight magazines to encourage visitors to plan a visit to the airport duty/tax-free shops on their arrival. Passengers are able to purchase goods that will be held for them until their departure.

Bus shelters, billboards and car parking barriers: All three of these facilities can be used to target different consumer groups. As an increasing number of airports are open 24 hours a day, transport activity between terminals is almost continuous. Bus shelters are considered an effective means of targeting staff travelling to and from work, while billboards are placed along roads and parking areas to target consumers who arrive by car. Paris airport duty/tax-free shops use bus shelters, billboards and poster sites to specifically advertise airport retailing and emphasise the range and quality of goods in their shops. Similarly, car parking barriers are now used in some airports to display company logos and product information.

Local media: This admittedly catch-all category includes advertising on local TV and radio, in local newspapers, and on the sides of taxis and buses. Non-travelling consumers living within the airport catchment area can be encouraged to use the landside facilities. Alternatively, press and radio may be used to highlight promotions and exclusive offers available to the travelling public. For example, special editions of best selling titles are often on sale in airports before being released in the home country with local media being used to communicate the exclusivity of these products.

Communication within the Airport

Successful retail advertising within the airport has to contend with flight information as well as overcoming the anxiety of many passengers. Nevertheless there are a number of media that both the airport authorities as well as retailers can use to communicate with consumers. These include;

Printed material: many airport operators place restrictions on the amount of printed material that can be distributed through the airport. This is to avoid inundating passengers with excess information that they will be unable to assimilate over the period they are in the airport. At the same time, airport authorities must ensure that their concessionaires have the opportunity to fully inform consumers on the products they have available. One solution to this issue has been the publication of in-terminal brochures by the airport operator advertising the full range of retail products on offer. In most cases these brochures are funded by supplier advertising and compiled jointly by the airport authority and the concessionaires. Typically, as in the case of Milan airport, the brochure will detail all the special offers in each product category and be published on a bi-monthly or six-monthly cycle.

An alternative approach adopted by some airports has been to produce smaller, more targeted brochures that focus upon special offers, specific holidays or public events. Sky Shops at Brussels airport distribute special offer leaflets to all passengers on entry to the airside facilities. These single sheet flyers focus on one or two products and may offer discounts or a free gift with purchase. Aer Rianta has developed over seventy customised brochures whose target audiences include rugby and football fans as well as more general Christmas, Easter and holiday travellers.

Most operators also produce their own magazines and terminal guides that are available to customers throughout the airport. For example, the magazine produced by Copenhagen airport includes special features, news of local events and activities as well as information on the range of retail services available within the airport. The publication also includes special offers from the duty/tax-free shops. Both retailers and suppliers have the option of placing advertisements in the magazine. As a way of increasing the traffic flow in the landside facilities and targeting non-travelling consumers, both operators and retailers have also increased the amount of advertising placed in staff magazines.

Airport television has become an accepted feature of many airports. Often providing 24 hour news, film and feature coverage, it also offers the opportunity for retailers to advertise their products through commercials and infomercials. The TV system at Rome airport also has a channel that provides weather forecasts for the principal regions of the world. Airport TV has the advantage of being able to penetrate all areas of the terminal from arrivals, to restaurants, to departure gates. Because of the expertise required to provide such a service, TV production is typically contracted out by the airport authority to a third party communications specialist. Usually this will be in the form of an advertising concession, which will be expected to earn revenue for the airport. In many airports, because of the significant investment needed to install these systems, the concession may have a longer than usual licence period to allow for full depreciation of the capital involved, with the hardware reverting to the airport on the expiry of the concession.

Lightboxes: As their name suggests, lightboxes allow organisations to display illuminated advertisements. Lightboxes tend to be a common method of in-terminal as well as in-store communication. They provide the opportunity for target marketing, by allowing the company to advertise in the most appropriate part of the airport (arrivals/departures etc.). In many airports, for example Frankfurt and Heathrow, retailers use lightboxes sited on the main routes to the terminals from the airport car parks, and in the corridors between the aircraft gates and the main departure shopping areas.

Promotional points: A number of airports have promotional points; these are locations within the terminal building that allow specific products to be advertised, often in the form of a display. Companies using promotional points have a degree of discretion over how they use the area and may decide to distribute flyers, hand out free samples or display their product. Increasingly, advertisers are also integrating multi-media technology with product displays in order to involve customers in the assimilation of product information. These promotional points are often used in conjunction with other forms of media within the airport, such as an advertisement in the airport magazine, a series of lightboxes in the terminal or an outdoor poster site in the airport environs. In some airports, especially in the Middle East, a car on display at the promotional point is also the prize in a limited sweepstake. The airport operates the car draw as a revenue earner and sells tickets at up to $100 each, with a maximum of a thousand tickets for sale. The airport will deliver the car to the winner anywhere in the world.

Show-cases and advertising pillars: Show-cases allow full visibility of the display from all sides and are often used to focus the consumer's attention upon a specific product or service. They are extensively used within the terminal building and may comprise anything from a static display to a revolving door insert. Show-cases are often used by airport retailers to illustrate the diversity of their product range. In addition to being placed within the terminal building, show-cases may be located in hotels, local restaurants and in tourist information centres. Advertising pillars, by contrast, are double-sided stainless steel structures that are often placed in the proximity of waiting passengers. They will have advertisements attached to each side and may be back-illuminated as well as providing printed media for information.

Luggage trolleys, carrier bags and conveyor belts: Because of the sheer volume of passengers travelling through an airport, trolleys, carrier bags and belts represent valid advertising media. Schiphol, for example, has 7000 luggage trolleys and distributes five million carrier bags annually. Branded goods suppliers, airlines, mercantile card issuers and car rental agencies are among those firms who regularly use this form of media. Many airports use the duty/tax-free shopping bags as important image generators, as travellers tend to re-use the carriers when back home (possibly for status reasons and to show the extent of their travel). For example, the yellow, black and red design of the Schiphol shopping bag is probably recognised the world over. See Table 4.2 for typical values of the advertising rates and specifications for Schiphol airport.

Table 4.2　　*Examples of advertising rates and specifications for Schiphol airport, 1996 (Dutch guilders ex VAT).*

Display type	Cost	Specifications
Lightbox	30,000–70,000 p.a.	One 50 × 100 cm transparency
Promotional point	12,000–27,000 per month	500 × 500 cm display area
Showcase	9000–65,000 p.a.	Average size 70 × 70 × 140 cm
Conveyor belt	72,500	One visual display on nine belts for twelve months
Catalogue	25,000 per issue	20 × 23 cm on back cover
Carrier bags	235 per 100	Minimum order 200,000
Luggage trolleys	155 per trolley p.a.	Minimum 500 trolleys
Schiphol TV	8000 per month	20-second spot broadcast 12 × per day

Source:　Schiphol Airport Authority (1996).

In-Store Communication

This remains the sole responsibility of the retailer and only in exceptional circumstances will the airport operator play any role in in-store communications. At certain times of the year, such as the opening of the main holiday season or in the weeks prior to Christmas, the airport authority may organise sales promotions throughout the terminal, focusing on specific products categories such as perfume or malt whisky. In such cases the shops would be fully integrated into the promotional strategy.

In addition to the principles of layout and design described in Chapter 3, the primary ways in which retailers communicate with their customers are through signage, light boxes and merchandise displays. Joint supplier/retailer promotions are also held on a regular basis where specific products will be marketed. Suppliers often provide promotional assistance either through a financial contribution or the provision of extra staff. The retailer will provide the promotional material, such as in-store posters, shelf-talkers, gift-wrapping services and special display units.

Only if the retailer or supplier wishes to expand the promotion outside the store is there a need to involve the airport authority. Such events however are relatively common and an airport wide promotion often coincides with a supplier's national advertising promotion. This allows the airport to 'piggyback' its own campaign upon that of the suppliers to ensure the greatest utilisation of marketing investment and the maximum customer exposure.

Airport retailers apply many of the promotional and brand marketing techniques used by the multiple retail chains. Trained consultants, often funded by the main suppliers, are used to promote sales and provide product information to customers. These are used to considerable effect in the perfumery and cosmetic areas where the product ranges are complex and extensive. New products and line extensions arrive on a regular basis and consultants may provide advice on the suitability of skin care products as well as a colour advisory service. In Madrid and Barcelona airports the perfume shops are completely self-service, although consultants are positioned at many of the main floor merchandise units to assist customers.

Many airport retailers regularly use specific promotions to cross-sell products. Sky Shops at Brussels airport have a policy of 'rolling promotions' which combines different merchandise assortments. For example, one initiative entitled 'Tie One On' raised sales by 25%. Customers who purchased any bottle of whisky could also purchase a silk tie from a range of over two hundred for only a quarter of the normal price. The same ties were also on sale at the normal price so customers were able to determine for themselves the value of the promotion. The offer was communicated to potential customers through an integrated campaign using show-cases, lightboxes, printed flyers, shelf 'wobblers' and in-store personnel promoting directly to the customers.

When an existing supplier wishes to advertise a specific brand within the terminal, the airport advertising concessionaire is required to keep the airport authority and the retailer informed. (Most concession agreements contain a clause that allows the airport to approve all potential advertisers.) This ensures that a consistent marketing message can be communicated within the store as well as within the airport. One issue that can arise is a supplier advertising a specific product not stocked within the airport. This can lead to a number of customer complaints about availability and an overall decline in service delivery.

Delivery of Value

The final stage of the exchange process is perhaps the most difficult to achieve with any degree of consistency. The effort of retailers to understand their customers, create the right product assortment and communicate their offer, is often undermined by a poor service delivery. Service failure occurs for a number of reasons, for example the unavailability of product, poor product knowledge of the staff, lack of customer service skills or absence of after-sales service.

For both the retailer and the operator, the aim is to ensure that the product is always available and that the service is consistent with the airport's image and the values that the customer holds. Two issues stem from this strategy. First, the importance of logistics and supply chain management in ensuring product availability. The old adage of 'getting the right goods in the right place at the right time in the right quantities in the right condition' still remains applicable in the highly competitive travel industry. In Chapter 5, the contribution that logistics makes to the delivery of value and the exchange process is examined in detail.

The second issue is human resource related. As a means of differentiating the service delivery within an airport there exists a need for competent, well-trained and motivated staff. The issues surrounding the management of people in airports and the methods employed to ensure a consistent delivery of value will be explored in detail in Chapter 6.

One way of conceptualising the delivery of value is as a series of dis-crete yet interrelated stages operating within the exchange relationship. Sparks (1991), for example, divided the delivery process into three discrete periods. The *pre-transaction, transaction* and *post-transaction* stages. The *pre-transaction stage* often occurs away from the retail store. Within the context of airport retailing, the pre-transaction stage may include personal queries about product availability, price comparis-ons and information on duty/tax-free allowances. In order to aid value delivery at this stage in the exchange process, BAA for example has a free information shopping line and a terminal help desk to assist with customer enquiries. Dublin airport operates a customer service desk that is fully manned whenever the shops are open. Birmingham airport operates a tele-sales unit at the airport which sells holidays directly to customers who telephone. Information on what products are available at the airport shops and the latest special offers is also avail-able.

The *transaction stage* relates to the actual exchange itself. In order to ensure that consumers receive a level of consumer service that is commensurate with the airport's market positioning, operators have introduced a number of service provisions to their retail offer. Issues of importance here may include sales staff skills, product knowledge and check-out queue length. Some retailers have even undertaken to carry any heavy items up to the departure gate for customers. In order to con-tend with the time pressures many passengers experience, a number of airports provide extra staff at busy periods to pack customer purchases at the check-out.

Laser scanning, rapid credit card approval, pre-wrap of best selling gifts, speed registers, individual product keys on tills and packing at check-outs, all speed up the transaction stage and contribute to a consumer's perception of the delivery process. An important factor in ensuring a smooth exchange process is the presence of a floor or duty manager who can manage queues and keep a constant flow of customers to all registers. The transaction process can also be aided by the ability of customers to pay at any register in the store and not be confined to a par-ticular till. This facility can be of critical importance at peak times. Many passengers will not queue because of a fear of missing their flight and simply walk away, resulting in a lost sale and poor delivery of value.

The *post-transaction stage* often relates to product returns and complaint handling procedures. One reason for the unwillingness of consumers to buy electrical goods in an airport has been the per-ceived difficulties in returning the product should a fault occur. To help overcome this, BAA has what it describes as a 'no quibble' guarantee return policy. This means that it is willing to take any product back without argument. Moreover, it allows goods to be returned by post for a full cash refund with the company paying the postage. Other airports, such as Dublin, have arranged with their suppliers to exchange any faulty product rather than go through a repair process. Customers returning defective goods are automatically given a new item or their money back.

The delivery of value through the three transaction stages should theoretically become a seamless exercise. The difficulty lies in ensuring that this objective is actually met and consumer-related values are consistently delivered. Service and delivery breakdowns are unfortunately all too common in retailing, and airports remain no exception. One approach to helping to reduce service failures is by conceptualising the delivery of value through Thomas's (1987) interfaces model. Thomas (1987) showed that delivery breakdowns occur through an interactive failure between management, staff, the retail system and consumers (Figure 4.3). Within retailing there are six primary interfaces. These are between:

- management and the customer;
- staff and the customer;
- management and the staff;
- customer and the system;
- management and the system;
- staff and the system.

Systems in this context are defined as the way in which the product or service is delivered. This represents a wide array of influential factors over which the provider needs to maintain control. The speed with which an order is taken, the queuing time, the provision of order status information and delivery times are all examples of how the system can influence the exchange process. In the context of airside retailing, this situation is compounded by the process that an individual has to go through before purchasing. Check-in, security and passport control are all systematic functions over which the retailer and the airport operator may have little control, yet these can influence the relationship with the customer. If airports are to provide a total quality proposition for their passengers, full co-ordination between all stages of the airport service chain must be ensured. This includes service providers not under the authorities' direct control, such as bureau de change, catering and car hire as well as retail concessions. There is a requirement therefore for both airport staff and management to be fully conversant with the control systems within an

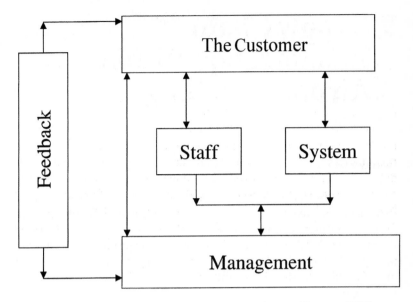

Figure 4.3 *Interface model of customer care (source: Thomas, 1987).*

airport and to reduce the complexity of the systems over which they have responsibility.

The relationship between staff and management also remains fundamentally important in the delivery of value and highlights the importance of good human resource management. Management needs to develop good working relations with staff, as inadequately trained or poorly motivated individuals can undermine the exchange process and lead to poor delivery. Similarly the relationship between customers, staff and management is fundamental to a continued exchange process. As Figure 4.2 illustrated, the model is iterative and not linear, and the objective for any retailer is to maintain competitive advantage through the continual delivery of value.

5 Supply Chain Relationships within Airport Retailing

Introduction

The aim of this chapter is to examine the structure and operation of the supply chain in airport retailing. It will examine the types of relationships that exist between retailers, airport operators and suppliers, and consider the methods employed to source, buy and supply product. The typology of airport retailers developed in Chapter 2 will provide a framework for understanding the similarities and differences that exist between operational structures.

The composition and operation of the supply chain within airport retailing can be seen both to mirror aspects of other retail sectors as well as to display some unique features. This chapter details contemporary supply chain developments before examining the factors that specifically influence buyer/supplier relationships within airport retailing. The process of negotiation and the responsibility for product availability and service levels within an airport will vary depending upon the contractual approach adopted by the airport.

Power in the Supply Chain

The supply chain and the interaction between supplier and retailer have traditionally been predicated upon the notion of some form of power relationship. How this is characterised and the way in which it is manifest is the subject of considerable academic debate. Much of the literature concentrates upon trying to develop a means of identifying and measuring power. Gaski (1984), for example, maintains that there remains no adequate definition of what constitutes the key elements of power.

Among the earliest views on power relationships within supply channels are French and Raven (1959) and Emerson (1962). French and Raven (1959) identified five sources of power; these were reward, expert, legitimate, referent and coercive. The ability of one party to provide rewards, expertise, legitimate behaviour, identify inter-party association or punish another represents the basis of their theory. Emerson viewed retail–supplier relations in terms of dependence. That is, one firm's dependence on another is defined by the extent to which one firm has to rely upon the other in order to achieve its goals and the availability of alternative methods by which those objectives can be reached. The French and Raven theory was later amended by Hunt and Nevin (1974) who identified the difficulties in differentiating between the first four types of power. Accordingly they developed a dual model of coercive and non-coercive power. The former provides rewards while the latter seeks to punish. A number of studies have sought to examine the supply chain in terms of this dichotomy. Rushton (1982), for example, noted that coercive power increased channel conflict while non-coercive tended to decrease it. Similarly Gaski and Nevin (1985) found that the presence and use of coercive power have a negative effect upon channel member relationships, while the presence of non-coercive power has a positive effect.

Bowlby and Foord (1995) maintain that many researchers fail to distinguish between exercised and unexercised power. Ogbonna and Wilkinson (1996) maintain that exercised power exists when one party undertakes a course of action that it otherwise would not have done. The power to achieve this may be based upon the degree to which one party is dependent upon the other for profit or sales, the availability of substitutes or the costs of switching to another relationship. Unexercised power, in contrast, refers to one party having the potential capacity to alter the behaviour of another party. Despite being difficult to conceptualise, an understanding of such power relations remains fundamental to understanding both operational efficiencies and conflicts within the airport supply chain.

Relationships within the Retail Supply Chain

The extent to which power manifests itself within the supply chain will be a function of the levels of competitiveness within the industry, the particular product characteristics, the size of the firms operating, as well as external influences such as legal constraints upon the relationship. Although very general, it is possible to understand supply chain relationships in the distributive trades in the 1950s and 1960s in the context of Emerson's (1962) dependency theory. The fragmented nature of the sector meant few retailers could exercise significant purchasing power and the relationship was one of retailer dependence upon manufacturer brands. While it remains accurate to suggest that manufacturers continue to exercise significant economic power in the marketplace, the 1980s and 1990s witnessed a fundamental change within the European retail sector. The asymmetrical power relationship that existed between retailer and supplier underwent radical readjustment.

In sectors such as grocery and clothing this led to a restructuring of the supply chain and a shift in the balance of power away from the manufacturer. Ogbonna and Wilkinson (1996), for example, highlight a complex set of interrelated power relationships that characterise the UK grocery sector. These range from mutual dependence based upon countervailing power between the major manufacturers and retailers, to smaller retailers developing strategic alliances with secondary suppliers.

The concentration of market share has become a feature of the retail sector. This in turn has had implications for a number of manufacturers. For example, suppliers are constrained in their delivery base and are compelled to negotiate with a limited number of multiple outlet retailers who used procurement as a competitive weapon (Baden-Fuller, 1986; Wrigley, 1993, 1994, 1996; Bowlby and Foord, 1995). The power that individual retailers now exert on the marketplace has been further reinforced through a strategy of centralisation. A number of retailers have shifted most of their decision making away from the stores and concentrated it at the head office (Freathy and Sparks, 1995). Stores are therefore free to focus upon improving sales while other, more strategic functions are the responsibility of specialist personnel. This separation of conception from execution has provided retailers with a number of benefits. For example, product purchasing is typically undertaken through a limited number of buying points at head office, which not only increases efficiency and improves margin but also ensures a greater degree of centralised control.

The negotiating power of domestic retailers has also been strengthened by the growth of own label product. Supplying goods under the retailer's own brand name has a number of advantages. It not only provides the retailer with an improved gross margin (10–15% higher than on manufacturer brands) (Siguaw and Hoffman, 1995), but also gives the retailer increased control over the characteristics of the product in the development stage. Retailers have therefore made positive attempts to increase their portfolio of own brands. Ogbonna and Wilkinson (1996), for example, note that in the UK, 54% of Tesco's sales are own label.

Fernie (1995) maintains that the adoption of supply chain management techniques varies significantly both within and between countries. For example, in grocery retailing, the UK is considered to be much further down the experience curve in logistics than other European countries. This is primarily due to its innovative use of technology and the collaborative relationships that exist between retailers and suppliers.

Nevertheless the supply chain throughout Europe has undergone radical change as retailers have sought to improve efficiencies through the better control of their stock and product flows. Responsibility for holding stock has shifted from retailer to manufacturer. Suppliers themselves often hold less stock than previously. Cooper *et al.* (1994) identify three factors that have reshaped the logistics function and created a much greater transparency within the supply chain: these are investment in technology, reduction in storage space and third party contracting.

Retail investment in technology has been significant. In addition to providing increased customer service and management information data, systems such as electronic point of sale (EPOS) and sales based ordering (SBO) have provided organisations with improved operational effectiveness. Fernie (1995) notes that electronic data interchange (EDI) in particular remains a central factor in both promoting collaborative arrangements and improving distribution efficiencies. Smith and Sparks (1993), for example, highlight how technology has allowed retailers to improve communication within the supply chain, tailor their product offer more closely to consumer demand and improve the accuracy of their supplier payment systems.

Investment in technology has been accompanied by changes in the methods of product distribution used by retailers. In a number of European countries, such as the UK and The Netherlands, there has been a movement away from suppliers delivering direct to store in favour of a system of centralised warehousing. This has allowed quantity discounts, better control of inventory, reduced shrinkage and breakage, and the opportunity for mechanisation. Major retailers may channel up to 90% of their stock through centralised warehousing (Fernie, 1992). Such a situation is not common to all European countries. In France and Spain, for example, the existence of the hypermarket and the small number of widely dispersed sites make the development of centralised warehousing infeasible. At the store level, this has the effect of reducing the amount of storage space required, increasing the net sales area and reducing the amount of capital tied up in buffer stock.

The use of third party contractors has been another significant development within the supply chain. Specialist contractors have been used to cover a wide variety of supply chain activities including warehouse management, control of information systems, fleet operations and order processing. The use of composite vehicles also reduces the number of deliveries made direct to store and provides economies of scale. The benefits derived from this strategy include a better financial return on capital employed, access to the most up-to-date technology, improved productivity and a more responsive customer service provision.

Types of Channel

The developments described above have had the effect of fundamentally restructuring the supply chain. It therefore remains possible to identify two polarised structures that characterise working relationships. Lusch (1982) saw the *conventional* channel as a loosely aligned, fragmented series of paired relationships (dyads) between different members of the supply chain, for example manufacturer–wholesaler, wholesaler–retailer, retailer–consumer. The focus is upon controlling and managing the relationships in these joint interactions. The relationship is seen as one based upon market exchange principals. For example, buyers and suppliers negotiate over prices, volumes and margins. While mutual trust may be absent in the initial stages of the relationship, interdependence does exist and, depending upon the working relationship developed between the two parties, can increase over time (Dwyer *et al.*, 1987). However, such an approach is seen as an inefficient method of distributing product and remains characterised by high levels of distrust, rivalry and secrecy.

The competitive nature of the retail sector has led to a reassessment of this strategy. In an attempt to improve margins and develop a sustainable competitive advantage, a number of retail organisations have sought to improve the working relationships they have with their suppliers. By moving away from a conflictual approach, emphasis has been placed upon improving communication, sharing information and co-operating more fully. Such relationships are long term, encourage discussion and negotiation, and cover a wide range of supply chain issues including product characteristics, quality management and new product development. This vertical marketing system (VMS) attempts to build a degree of loyalty and is premised on the notions of relationships and co-operation. While many different forms of vertical marketing system exist, two particular types have attracted the greatest academic interest:

Corporate, or vertically integrated systems: These exist when a company owns more than one part of the supply chain. Typically the retailing and manufacturing elements of the operation will be owned by the same organisation. In addition they may control the distribution and even the primary production facilities. Oil companies with their ownership of rigs, pipelines, refineries, tankers and stations are often cited as examples of organisations with highly integrated systems. The benefits of a vertically integrated structure stem from the greater power that can be exerted in the marketplace. The risk of supplier change is reduced while greater control over product standards and the retail offer can be exercised. However, corporate systems are complex and difficult to manage and, in periods of economic downturn, companies run the risk of paying fixed costs on redundant capacity. While vertical integration was more common in retailing in the 1960s and 1970s, its popularity as a method of channel management has waned since the 1980s.

Contractual or vertically co-ordinated systems: Unlike corporate channels, the supply chain comprises a number of separate organisations who are bound together in a series of contractual relationships. The extent to which this form of agreement operates within Europe varies dramatically. Fernie (1995), for example, maintains that the attitudes toward co-operation within Europe vary markedly, with retailers in Spain and Italy voicing distrust toward such a strategy. Since the 1980s vertically co-ordinated systems have grown in popularity and are often championed by one (often the most dominant) member of the channel. Contractual agreements place obligations upon all parties within the supply chain. Perhaps reflecting their increased market power, the initiative for the development of vertically co-ordinated systems has been primarily the prerogative of the retail sector. Such agreements place a number of obligations upon all parties including delivery standards, support services, marketing information advice and production assistance. The extent to which contractual systems operate effectively is a function of the market power exercised by individual organisations within the chain and the degree to which this power is viewed as legitimate. Fernie (1995) notes that many manufacturers were initially hostile to the notion of partnership, however the growth of private label provided the impetus for closer contractual ties. Companies who exercise considerable influence within the marketplace and wish to develop this form of relationship remain most likely to achieve a positive outcome.

While vertical systems have been advocated on the basis of productivity and efficiency, no supply chain remains devoid of conflict. Lusch (1982) identifies three sources of conflict within the supply chain. Perceptual incongruity occurs when members of the supply chain have different views on its operational effectiveness; for example, the quality of the delivered merchandise may be perceived differently between the retailer and the supplier. Similarly the expected level of product support provided by the supplier may be at odds with what the retailer expects. These differing perspectives held by channel members represent major sources of conflict within the supply chain. The second source of conflict occurs when channel members have incompatible goals. The areas where disagreements arise in this context will be widespread and may revolve around issues of capital investment, strategic timings and the provision of inter-organisational support. The final area of conflict occurs where responsibility for decision making remains poorly defined. This 'domain dissensus' is often the result of poor planning and a failure to agree specific tasks. Securing agreement on each member's role is vital for a VMS system to work effectively. If any part of the supply chain process remains unaccountable then the holistic ethos of total responsibility becomes redundant and the channel will revert to a contractual, exchange-based relationship.

An allied although different form of relationship that has emerged also merits discussion. Buying groups are different from the contractual forms of relationship described above in that they are inter-organisational alliances of European retailers who have sought to co-operate together on operational and strategic issues. The rationale behind these associations may be a defensive reaction to the practice of dual distribution or a strategy to counter manufacturers' consolidation of market share through merger, acquisition and take-over (Robinson and Clarke-Hill, 1995). Alternatively it may be offensive and seek to improve competitiveness through inter-party co-operation. Such alliances however primarily seek to improve the efficiency and effectiveness of the members through co-ordinated supplier selection, common pricing, discount sharing and price bench-marking.

Supply Chain Operations within Airport Retailing

There has been a movement throughout Europe to centralise many of the strategic functions that have traditionally been the responsibility of the store. The introduction of scanning and retail information systems has been a key driver in this process and has allowed the development of efficient information led distribution. To date, however, many of the benefits of new technology in airport retailing have focused upon speedy store throughput and handling, "rather than to address those issues of integrating the independent elements in the supply chain to maximise service level and minimise costs in order to gain a competitive advantage" (Slater, 1990, p. 149)

The ability to integrate retail information systems with passenger and aircraft traffic flow data allows the possibility of creating a centralised system of distribution as well as developing more accurate merchandise forecasts.

The extent to which airports mirror supply chain operations in other retail sectors varies according to the type, size, structure and ownership of the airport. There remains no single approach to understanding the supply chain process. The typology of airports identified in Chapter 2 can be used to help conceptualise operations, however the examples given below represent a broad aggregation of the methods employed and remain an illustrative rather than an absolute definition.

There remain two features of airport retailing that distinguish it from its domestic counterpart. First, the demand for specific products within an airport is a function of the volumes and types of passengers travelling rather than the time of year. While promotional initiatives have the potential to influence purchasing behaviour, demand remains stable when compared to the high street. Duty and tax free is viewed as a volume business with sales patterns over the year remaining relatively constant. One exception to this is perfumery where gauging sales demand remains more difficult to predict (although the control and availability of stock has improved through the introduction of EPOS and other forms of electronic information systems). A second feature of airport retailing is the disproportionately high concentration of sales in specific product areas. As Chapter 2 illustrated, the sales of products such as liquor, tobacco, perfume and confectionery dominate airport retailing and represent the primary sources of revenue for the airport operator. The efficiency of the supply chain in these product categories therefore remains critical to the overall success of the airport commercial strategy.

The Supply Chain Process

The basic supply chain within airport retailing may be broken down into four stages: the ordering process; the supply of the product; warehousing, storage and inventory management; and the movement of product from warehouse to the point of sale.

Ordering process

Because of the high volumes of tobacco and liquor sold in duty free, product demand is reviewed and ordered by the retailer on a weekly basis. The lead time from ordering to delivery will depend upon the product and the source of the supply, and may range from 24 hours to several weeks. Timings are agreed between the parties during the negotiation stage. Tax free and other merchandise tend to have a longer order cycle time. While it is standard to review demand for these products on a weekly basis, delivery again tends to vary. For example, perfume may operate on a 3–4 week delivery cycle while fashion goods and crystal may take 2–3 months. When placing an order for merchandise, the retailer will typically provide the following information for the supplier:

- delivery date due;
- name and address of supplier;
- date;
- product description;
- quantity ordered;
- pre-arranged cost price per unit;
- freight and carriage costs;
- terms of payment;
- special terms and discounts;
- price look up number (PLU) for each product;
- the total amount of the order at cost.

Within the airport industry as a whole there is only limited evidence of technological innovation. While there remain some notable exceptions to this, for example, Heinemann, systems such as electronic data interchange (EDI) and JIT are conspicuously absent. Supplier replenishment and the majority of orders are manually placed with the supplier. In addition to being a slower and less efficient method of ordering product, there remains a higher likelihood of delivery and receipt error. This in turn has implications for the retailer as the majority of stock is not supplied on a sale or return basis to airports.

Supply of the product

The method of delivery of the product to the warehouse will depend upon the range of merchandise stocked and whether the buyer has a centralised or decentralised warehouse operation. Customs and Excise maintain close control over all duty-free products that are moved between supplier and retailer. In some instances the supplier will have to seek approval from the Customs for its shipment routes. The majority of duty-free products are delivered on pallets directly from the manufacturer's bonded warehouse to the retailer. Increasingly, merchandise is bulk packaged to the buyer's own specifications, which may include the size, style and colour. It is also expected that merchandise should be pre-priced and bar-coded by the time it arrives at the warehouse.

Warehousing, storage and inventory management

One of the greatest difficulties in warehousing duty-free products is coping with the high volume demand. Because of the high rentals charged for space by the airport operators, some retailers have chosen to establish warehouse facilities away from the airport. One perfume and cosmetics retailer estimated that outside the airport, rents were ten times cheaper for warehouse space and eight times cheaper for office space. Such a differential has obvious logistical implications, for example the perfume and cosmetics retailer had calculated that it was more economical to make three, eight kilometre return journeys a day rather than lease premises from the airport authority.

As both duty-free shops and warehouses are designated bonded areas by Customs and Excise, then this results in an extra layer of accounting and inventory management for the retailer. Under bond regulations all goods must remain in unopened cases while in the warehouse. For customs purposes, excisable products are classified by product group. Groups include whiskies, brandies, tobacco, cigars, wines and champagnes. Customs may also require a further classification by size of unit. In the case of alcohol this would include litre, 70 cl and 50 cl sizes. While the investment in EPOS systems has aided the control and accountability of retail stock, the level of accuracy required across merchandise categories is arguably higher in airports than in high street stores.

Deliveries from the supplier's bonded warehouse to the airport's bond are usually accompanied by a Customs and Excise warrant. This details the contents of the consignment and is used as a control check by local customs. The role of the Customs and Excise in this respect is to ensure that whatever stock comes into a duty-free warehouse equates exactly to sales plus stock on hand. Any subsequent inaccuracies in inventory levels can incur duty and tax fines by the customs at penal rates. This often means that the product with the highest excise rate is chosen and applied to all merchandise variances.

Once the supplier has delivered the product, it is checked through the goods inwards department. The congestion created by warehouse deliveries and the outbound transportation of product to the store has led some retailers to impose delivery schedules upon suppliers. On delivery at the warehouse, all product items are checked against the official company order and the supplier's delivery docket. Any differences are brought to the attention of the relevant buyer, the supplier and the company's own accounts. The retailer then provides the supplier with a signed proof of delivery (POD) notice for its records. On receipt of the delivery, Goods Inwards Notices (GINs) are raised by the retailer. A GIN will be the official company record of the receipt of the product from the supplier and specific information is therefore required. A GIN will typically have the following information:

- date of goods received;
- order reference number;
- delivery docket reference number;
- PLU number on line;
- description by line;
- quantity received by line.

Movement of product from warehouse to point of sale

Among the largest duty-free retailers there is only limited evidence of third parties being subcontracted to move product from the retail warehouse to the store. The majority of retailers use their own transport facilities for composite delivery. In contrast, the duty-free provision at smaller regional airports is often not large enough to justify direct supplier delivery or investment in separate warehousing facilities. In these circumstances composite warehouses operated by independent third parties act as both a wholesaler and direct distributor to store.

The supply of product from the warehouse to the store transfers goods from one bonded area to another. It is therefore done on the basis of an official requisition that may be checked by Customs and Excise at any time. A requisition typically states:

- the date;
- the PLU number;
- the location;
- the quantity required
- the description of the product.

On receipt of the goods at the retail outlet, the quantities, the PLU and the prices are checked against the requisition and all products are checked for damage. If accurate they are accepted by the store. Any breakages have to be officially accounted for. In the case of duty free, the damaged items have to be handed over to the customs for disposal; for example, the retailer may be required to provide the unopened neck of a liquor bottle in order to demonstrate that the product has been damaged. If it is not possible to provide evidence that the product has been broken, they will appear as stock shortages and be liable for VAT and duty charges.

The Buying and Merchandising Process

Buying remains a critical element in the supply chain process and one controversy that exists is whether a centralised system of buying provides greater economies of scale than a decentralised system. Where the airport retailer has responsibility for retail operations in both domestic and international markets then the correct choice of organisational structure becomes critical. A decentralised system of buying has the advantage of allowing each airport retail operation the autonomy and responsibility to control its own merchandise mix as well as the opportunity to develop its own marketing strategy. Moreover it creates a degree of accountability for each of the retail operations. There are also instances when a decentralised approach remains the only practical solution to retail operations. For example, joint overseas partnerships (a Type 4 operation) lend themselves toward a devolved structure. Attempts by some operators to centralise retail buying have led to difficulties in distribution, stock replenishment, the sourcing of local craft and souvenir merchandise, as well as legal disputes over exclusive distribution rights for major international brands.

Decentralising the buying function has a number of negative consequences however. First, there remains a duplication of effort both on the part of the airport operator as well as the suppliers; for example, in assessing the suitability of a new supplier, buyers often fail to visit the manufacturer on a group basis. In some instances therefore a supplier may have to repeat the initial evaluation process every time it wishes to supply an airport outlet controlled by the same authority. Secondly, a supplier's performance is often not monitored on a group wide basis. The difficulty of this situation is compounded when each airport uses its own generic supplier code and in some instances its own stock descriptors. This makes the comparison of operations such as stock flow and stock turn almost impossible to assess. Thirdly, there remains a duplication of administrative procedures and personnel. Each airport will have its own buyers, merchandisers and support staff. As well as increasing the layers of bureaucracy, such a system can add significantly to the overall running cost of the airport. Fourthly, negotiations with suppliers are regularly undertaken in isolation. This means that there is often little communication and co-operation between buyers from different airports on anything other than the largest volume products. The opportunity to attach group terms to a contract is therefore lost. It relies upon informal mechanisms or the good-will of the supplier to offer comparable prices across the group. Unsurprisingly this is often not the case and manufacturers will often seek to exploit the decentralised autonomy of each airport outlet.

The suppliers of the largest, most popular brands may insist that margins, marketing and advertising support for specific products are *not* agreed on a company wide basis. If the company has retail operations in a number of countries it may be under pressure to negotiate separately for each geographical region. One outcome of this is that the negotiated terms and conditions attached to each individual product may vary depending upon the location of the outlet. Ultimately this can lead to different stores within the same group paying different prices to suppliers and retailing the product at different prices to the consumer.

It is the role of the duty/tax-free operator to develop a buying structure that is able to cope with the industry's high volume demand. This structure may take many forms; for example, if the airport operator is responsible for the purchase and merchandise of product, then a separate retail function will be incorporated into the authority's overall commercial structure (Figure 5.1).

As Figure 5.1 illustrates, the airport management is responsible for the purchase and control of stock and operates under a formal bureaucratic structure. The Head of Commercial Operations has the overall responsibility for the performance of the company's retail activities. In addition to the tax and duty free, the remit may include any concessionaire retailing that the operator may wish to introduce into the terminal building.

Depending upon the size of the operation, responsibility for purchases may be split between the duty-free and tax-free functions with a manager responsible for each of these activities. The pattern of demand for alcohol and tobacco is both more predictable and stable, and the Manager of Duty Free, with the assistance of the product buyers, may take personal responsibility for negotiating with suppliers. In contrast, many of the products in Tax Free have a more complex demand pattern as well as more variable order cycle times. Consequently the Tax Free Manager may be aided by an Assistant Manager as well as having a team of buyers responsible for specific product categories such as Perfume, Electrical, Jewellery/Watches, Fashion and Souvenirs.

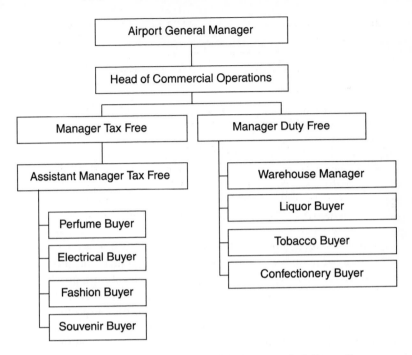

Figure 5.1 *Buying structure in an airport-managed retail operation.*

If the retailer is part of a national chain with high street stores then the airport operations will often be an adjunct to, or subsumed within, its main operations. For example, discussions with existing suppliers will take place as part of the general negotiation process. Products continue to be ordered centrally by buyers who are purchasing for the entire retail chain. While dedicated airport buyers exist, it is often their remit to focus upon extended ranges and bonding controls. Figure 5.2 illustrates a buying structure for a national retail chain with interests in the high street as well as the airport.

Power and Supplier Negotiation in the Supply Chain

No party in the supply chain is able to entirely dominate relations, and both retailers and suppliers are able to exert influence over the negotiation process. The basis of power for each party stems from different sources. For the retailer, the first and most obvious relates to the

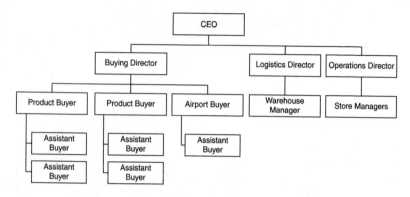

Figure 5.2   *Buying structure for a national retail company.*

volumes demanded. Stock turns remain high. For tax free, a stock turn of × 4 to × 6 per annum would be seen as normal, while in the duty free a stock turn as high as × 15 may be expected. Such volumes provide the retailer with a significant amount of negotiating leverage. A second and less obvious power source stems from the high-profile exposure that a product can receive if stocked in an airport. In addition to a sense of national pride that will influence a supplier's determination to be represented in its home market, manufacturers understand the value of profiling their products to an international community. This exposure is used both to establish a product's presence overseas and to generate demand in the domestic market. Airport retailing therefore acts both as a shop window and as an alternative distribution channel where suppliers can test-market their products.

One example of a retailer exerting its market power was the reaction of DFS to a decline in the Japanese tourism market. The consequent drop in revenue led the company to renegotiate a price reduction with all its suppliers. It aimed to maintain its current margins by expecting its suppliers to absorb the price differential (Duty Free Database, 1994).

Suppliers have not remained passive to the growing power of the retailer. In recognition of the importance of duty-free, the principal manufacturers have established specialist subsidiary companies to handle all their worldwide duty/tax-free sales. These organisations manage a portfolio of brands and represent the parent company in all aspects of supply. Retail concentration and the emergence of international airport retailers with centralised buying functions have also prompted a number of suppliers such as Allied Domecq, Rothmans and United Distillers to consider the introduction of global account managers responsible for individual clients. A strategy of centralisation allows bulk deliveries to be negotiated and more efficient distribution to be achieved. For the retailer, the global account model also offers much higher client service levels and the opportunity to exploit the full marketing potential of the supplier's brands. Many suppliers have not welcomed such a move, and have advocated a system of regional negotiations where retailer quantities (and consequently buying power) are less.

There has also been a strategy of supplier consolidation within the duty/tax-free industry. Major international brands have a tendency to change ownership on a regular basis. One outcome of this is that power has become concentrated among a limited number of multi-national corporations (Table 5.1). For example, United Distillers control five of the world's top twenty brands of spirits. Manufacturing companies have also sought to increase their power base by integrating forward into logistics and, in some instances, control over 80% of product movements.

Table 5.1 *Major leading suppliers and brand ownership, 1996.*

Supplier	Brand
Alcohol	
Allied Domecq	Ballantine's, Beefeater, Canadian Club, Carolans, Cockburns, Courvoisier, Laphroaig, Harvey's, Kahula, Teacher's, Tullamore Dew
United Distillers	Johnny Walker Red, Black, Blue, Gold and Premier label, Pimms No. 1, Asbach Uralt, Cardhu, Glen Ord, Dimple, Tanqueray.
Worlds Brands Duty Free (Pernod Ricard)	Cognac Bisquit, Cork Dry Gin, Huzzar Vodka, Middleton Malt, Pastis 51, Pernod, Powers Irish, Ricard, Suze, Wild Turkey
Tobacco	
BAT	Benson and Hedges, Capri, HB, Kent, Kool, Lucky Strike, State Express 555
Gallagher	Berkeley, Gallagher, Silk Cut, Sullivan, Hamlet, Condor, Old Holborn
Imperial Tobacco	Lambert and Butler, Regal, Superkings
Philip Morris	Marlboro, Eve, Merits, Lark, Parliament, Virginia
Rothmans	Cartier, Dunhill, Rothmans, Samson

International suppliers derive much of their power from the brand loyalty that they gain from their consumers. Creating a high profile marketing image gives the supplier scope to exert tight control over a range of operational areas. Perfume distributors, for example, demand quality fittings and displays, set retail prices relative to the domestic market and place limits on the types of advertising that the retailer can undertake. To help reinforce their market position, the perfume houses have been successful in securing the European Union Commission's support for the operation of a selective distribution system.

It would however be misleading to suggest that all relationships between retailer and supplier are conflictual in nature. Both parties have a number of common objectives. For example, both groups will wish to achieve a high level of customer satisfaction that in turn leads to sales growth and volume increases. Quality and reliability also remain particularly important factors because of the difficulties and expense of returning defective or below quality merchandise. Suppliers and buyers alike will seek to minimise reductions and mark-downs, especially given the statutory limitations on passenger purchase quantities imposed by most countries. The leading perfume houses, for example, do not allow discounting to clear slow or unwanted merchandise. They will exchange the stock for proven lines in order to protect their brand image and prevent the stock reaching the grey or parallel markets.

Negotiation with Suppliers

The relationship between the largest suppliers and retailers is symbiotic with each party relying on the other for the achievement of its own objectives. The increasing levels of consolidation within the duty/tax-free industry has led to a concentration of power and an increasingly significant role for those involved in the negotiation process. The process by which a company supplies an airport retailer is complex and will depend upon a number of factors, including the volumes demanded, warehouse availability and whether it is a new or existing supplier. Similarly the composition of the negotiating teams will vary depending upon the quantities demanded, the product specification and the supplier in question. The airport retailer may negotiate on a face-to-face basis with its large principal suppliers (mirroring a process common with retailers in other sectors). For its smaller suppliers, such face-to-face meetings may not be necessary and the retailer will negotiate through written communication. A duty-free airport operator can be expected to stock between 600 and 1000 items, and negotiate with anything between 70 and 90 duty-free suppliers (although in both the Tobacco and Alcohol markets there are approximately five primary suppliers). Perfume and tax-free merchandise are more diverse; Aer Rianta, for example, holds over 6000 different stock-keeping units at its Dublin store.

The criteria used for selecting a new supplier will depend upon the type of products being bought and the volumes demanded. They should however include:

- quality of product presentation and service;
- price and margin;
- reputation and brand profile on domestic markets;
- safety standards;
- financial standing;
- reliability and flexibility;
- production capacity and ability to respond to surges in demand;
- procurement and advertising support;
- innovativeness.

A new supplier may be drawn from a variety of sources. In addition to making direct contact with retail buyers, suppliers may be identified in trade magazines, trade fairs and on the basis of personal recommendations. A supplier wishing to offer a product to a retailer tends to make initial contact with the retail buyer. Initial discussions tend to be over the phone. In many instances the supplier can be told at this stage whether the product is suitable. Typical reasons for rejection will include its unsuitability to the existing product portfolio owing to its weight, its dimensions or its quality. Alternatively the supplier may be unable to provide the quantities demanded, meet the delivery times expected or produce the product within the margins set.

If the supplier is able to meet the retailer's production demands, they are required to submit a written proposal before being placed on a potential supplier list. The retailer then requests all available information on the product, including its composition, where the ingredients are sourced, its development history, other stockists and the retail price asked by the competition. In addition, a site visit may be required to examine the manufacturing facilities and a request made to speak to other stockists of the product. If the retailer feels that the supplier has satisfactorily met the required standards then a trial case or consignment may be requested and the rate of sales monitored. The success of this probationary period will dictate whether the product becomes part of the regular product mix stocked by the operator.

For existing suppliers, contract renegotiations are held annually. Depending upon the product category and the size of the supplier, such a meeting may take anything from one to three days. Buyers tend to work towards achieving an overall gross margin for the product group for which they are responsible. This provides them with a degree of flexibility to negotiate over specific products. In undertaking the negotiating process, the retail buyer must ensure that the margin and pricing strategy is comparative with the competition. A variety of market intelligence sources are used in order to gather the relevant information. This is supplied primarily by the company's own marketing department. Typically, a buyer entering negotiations should have an intimate knowledge of the market as well as information on previous order quantities and daily, weekly, monthly and total sales performances. As duty/tax-free sales are normally a function of the passenger traffic volumes in the airport, the buyer will also be expected to have detailed traffic statistics, forecasts and passenger profiles available. Cost price, retail price and gross profit levels should have been calculated and the buyer should be aware of the total retail value of the product. Such prior preparation is necessary given the levels of expenditure on product and the likely seniority of the supplier's negotiating team.

The market power that many duty/tax-free retailers now command, merits the attendance of a senior figure from the supplier's side. Typically, the person negotiating on behalf of the manufacturer will have responsibility for an entire region such as the UK or Western Europe. It would be unusual for any retailer involved in selling duty-free products to negotiate solely with an individual brand manager. The supplier would be expected to send a key account manager who would be able to negotiate on each of the company's product lines. Some suppliers have their sales and client service teams structured on an individual brand basis. Estée Lauder, the US perfumes and cosmetics corporation, for example, has a manager in overall control of the duty-free business in Europe. In addition, separate brands such as Clinique, Aramis and Estée Lauder each has a dedicated product manager. During negotiations each product range will be represented by a brand manager who will conduct detailed discussions on the individual brand portfolio. Overall control however will be maintained by the principal negotiator. Depending upon the volumes demanded, other suppliers may structure their sales team by geographic region or by industry sector. Specialist managers would be responsible for negotiating with airports, ferries and airlines.

During negotiations a number of issues will be discussed and detailed on paper. Of primary importance is the cost price of the product. It remains a central objective for the buyer to maintain a price advantage over duty/tax-paid retailers without loss of margin. In negotiating the cost price, the buyer must consider the overall price positioning strategy of the airport, the retail price levels at competitive airports and the maintenance of a price differential with the local market. If buying for the landside operation, costs and margins must allow the retail price to be set at a level equivalent to, or better than, the high street in order to maintain the airport's value image. Volumes and settlement terms, payment periods and discounts will also be agreed and finalised. Given the international profile of airports, suppliers may wish to improve a product's brand awareness and negotiate over display areas within the store. The perfume houses in particular attempt to specify the position of their products relative to the competition, for example Chanel insisting on being sited beside Lancôme, and Estée Lauder on being sited next to Dior. Prime merchandising slots tend to be allocated on the basis of supplier size and volumes demanded. (Such a strategy is often at odds with the merchandise and layout principles identified in Chapter 3 and remains a source of internal conflict between buyers and merchandise planners.)

In addition to the buying team, the marketing manager may attend the negotiations to discuss promotions and advertising support. Some such as the duty-free retailing division of KLM (the Dutch national airline) hold two separate sets of negotiations: the first discusses volumes and margins, the second considers promotional support. It is common practice in airport retailing to work to a promotional calendar that identifies specific events linked to the airport, such as international football games, seasonal periods such as Christmas and Easter, and special occasions such as Mother's Day and St Valentine's Day. It will be the function at this stage of the negotiation process to agree with suppliers their role and contribution to these events; for example, suppliers may be involved in undertaking a price promotion during a holiday weekend. The objective of such promotional activity may be to encourage a greater uptake of the brand by exposing it to an international audience or simply to increase the volumes by discounting. In either case the supplier would be expected to maintain the margin to the retailer and absorb the additional costs. The supplier would typically provide all signage and promotional material and, if extra personnel are required, then these will frequently be made available at no expense to the retailer. The retailer's role will often be minimal, consisting of the provision of space for a product demonstration and in some instances providing a stand and basic fixtures.

A number of large, internationally branded suppliers also utilise a system of agents or distributors who negotiate on the supplier's behalf.

In the majority of instances these agents have exclusive control over individual product portfolios for the duty/tax-free market. In addition they may also have the distribution and supply rights in the domestic market. Branded manufacturers such as Swatch, Fuji and Sony all distribute through agents. The use of the same agent to negotiate and distribute a portfolio of competitor's products remains less common, although not unknown. For example, in the tobacco industry there are instances of the manufacturers BAT, Philip Morris and RJ Reynolds all using the same agent to distribute their merchandise. The majority of wine supplied to duty free also comes through such agents, who act as a consolidator for a large number of small wine producers. Agents have their own warehouse and distribution facilities within the country they are supplying. Their value to the airport lies in their ability to split bulk and supply on an open-to-buy basis, therefore relieving the retailer or airport operator of the requirement to invest in large-scale bonded warehousing.

In addition to the yearly negotiations, suppliers will have formal scheduled meetings throughout the year with retailers. Again the frequency of these meetings will vary. Principal suppliers in duty free, for example, generally consider it necessary to meet between 3 and 4 times a year. The purposes of these visits are to discuss any operational issues that may have arisen over the trading period, monitor sales progress and exchange information on the wider duty-free market. Matters surrounding product facings, product layout and stock movements may also be subjects for discussion. Occasionally suppliers will use these visits to give advance information on new product developments, line extensions and special promotions or to share the results of market research. If a supplier negotiates and distributes through an agent then the retailer may expect a higher frequency of visit. This is primarily because the agent has direct responsibility for product supply and is financially accountable to the vendor. Airport retailers in general prefer to deal directly with suppliers rather than agents or distributors. The elimination of an entire stage in the supply chain allows for increased discretion over margins. As agents often promote their suppliers' products in markets other than duty and tax free, they very often lack detailed knowledge of the air industry and do not have the authority necessary to allow flexibility in price negotiations.

On the basis of supplier negotiations and internal information systems, a procurement strategy for day-to-day ordering and stock replenishment is devised. This is, in effect, a sales forecast that the company may use for budgetary purposes as well as facilitating the supplier's production plan. Table 5.2 presents a concessionaire's yearly buying plan for hand-made chocolates. This plan represents the broadest demand forecast, which in practice would be further disaggregated on a monthly and weekly basis.

Table 5.2   *Yearly buying plan for hand-made chocolates (£ Sterling).*

Supplier/ Product	Previous year's sales	+15.5% Growth[1]	+3% Inflation[2]	+2% Other factors = sales forecast	+1.75% Shrinkage = optimum inventory	Optimum inventory at cost
Serfs						
125 g Ballotin	23,873	27,573	28,400	28,968	29,474	17,684
250 g Ballotin	35,538	41,046	42,277	43,122	43,876	26,325
500 g Ballotin	30,788	35,560	32,626	37,358	38,011	22,807
125 g Flat	84,799	97,942	100,880	102,897	104,697	62,818
250 g Flat	82,239	94,986	97,835	99,791	101,537	60,922
500 g Flat	62,693	72,410	74,582	76,073	77,404	46,442
Cocoa						
125 g Flat	17,362	20,053	20,654	21,067	21,435	12,861
250 g Flat	32,443	37,471	38,595	39,366	40,054	24,032
500 g Flat	10,706	12,365	12,736	12,991	13,218	7,931
Juno						
8 oz Flat	10,305	11,902	12,259	12,504	12,722	7,633
190 g Ballotin	21,936	25,336	26,096	26,617	27,082	16,249
250 g Ballotin	24,165	27,910	28,747	29,321	29,834	17,900
350 g Ballotin	21,580	24,924	25,671	26,184	26,642	15,985
500 g Ballotin	15,613	18,033	18,573	18,944	19,275	11,565
100 g Flat	47,160	54,469	56,103	57,225	58,226	34,935
200 g Flat	38,438	44,395	45,726	46,640	47,456	28,473
Total					690,943	414,562

[1] Real growth is factored at 15.5% based upon international departing passenger traffic forecasts for the year (the example here takes growth as a constant, in practice this will vary by month).
[2] Inflation is factored at 3% based upon the Consumer Price Index.

Payment to suppliers will in almost all circumstances be made after a specified period of time. This period varies and will depend upon the volumes demanded and the power relationship between buyer and supplier. Open accounts with payment between 30 and 60 days after the invoice date are most common. Some suppliers may attempt to expedite payment by offering extra discounts for prompt settlements. A problem with international buying and supply is that both parties have to contend with changes in the exchange rate and the fact that negotiated prices may vary as a result of currency fluctuations.

There are a number of ways to overcome the potential effect that this variability can have on margins. The first is to agree an exchange rate on the day of negotiation. This allows accurate costings and financial forecasts to be made. Other retailers forward buy or sell currencies in order to hedge against exchange movements. While this method may avoid currency fluctuations related to specific events, it remains more speculative and carries a higher degree of risk. Some suppliers promote a contract arrangement with the retailer. This sets the exchange rate within an agreed band and will only change if there is a variation above or below a fixed percentage.

Payment itself is usually in the form of banker's draft and paid on the receipt of an invoice. International retailers with commercial operations in a number of countries will maintain a number of different currency accounts, which provides them with a greater degree of flexibility in paying suppliers and overcoming exchange rate fluctuations.

Evaluating the Success of the Buying Function

It remains in the interest of both the retail buyer and supplier to monitor closely the effectiveness of their supply chain strategies. For both parties, airside retailing represents a unique challenge in that the consumer does not have the opportunity to come back tomorrow. Suppliers therefore need to ensure that product quality is consistent and supply is regular. The retail buyer must seek to maximise retail sales through the correct choice of merchandise at a margin consistent with the company's objectives and in line with passenger expectations. Critical success factors from a buyer's perspective may be classed under two broad headings, as detailed below.

Effectiveness Factors

The first factor relates to the maximisation of sales potential. There remain a number of ratios that may be used to identify the success of a buying strategy. The two most popular indicators used by airport retailers are: *penetration rate*, which is the total number of transactions in a shop as a percentage of the total number of international departing passengers; and *average spend*, which is the total turnover divided by the total number of international departing passengers. These ratios are considered to be the drivers of duty/tax-free retailing and the operator may use these indices as a performance indicator when making comparisons with other airports. Both measures are typically monitored on a weekly basis and compared with the current budget and the previous year's performance. Both measures focus upon international departing passengers. Average customer spend is frequently disaggregated by route. This measure can indicate the extent to which customers are buying their full duty-free volume allowance. In order to maintain or increase average spend ratios, particularly in an environment where individual purchase volumes are fixed, some retailers set a minimum unit sale value for buyers.

Airports with a high percentage of domestic travellers, such as those in the USA, calculate the average value of the transaction as the total turnover divided by the total passenger throughput. As this calculation will include turnover at both airside and landside shops and for departing and arriving passengers, one may expect the value to be considerably lower.

Customer satisfaction also remains a factor against which the effectiveness of the buying plan may be evaluated. The repositioning of airports towards a higher quality retail environment needs to be reinforced through a product offer that complements this overall strategy. As illustrated in Chapter 4, airports devote considerable energies to understanding their consumers. One criterion for evaluating the success of the buying function would be through the perceptions held by customers of the merchandise range and the pricing policy of the airport relative to the domestic market. For example, consumers in an airport will often 'price validate' the merchandise on offer; that is, identify one or more key products with which they are familiar and are able to make a price comparison with a high street retailer or with another airport. If these items are considered expensive then this influences their perception of the whole of the retail offer. From a buyer's perspective, the aim is to maintain a pricing strategy that reinforces a perception of value. The buying and pricing policies of liquor and tobacco at Amsterdam Schiphol are designed to price-validate the rest of the retail offers in the airport's shopping centre.

The final 'effectiveness' criterion is the development of new consumer segments. This remains an important element in the maximisation of penetration rates. It is largely dependent upon buyers identifying new areas of merchandising that would appeal to non-purchasing passengers and possibly encourage existing customers to trade up or add on additional purchases. For example, the sale of children's confectionery in presentation packs has been successful in attracting passengers who previously did not purchase tax-free products.

Efficiency

This relates to the efficient control, handling and availability of merchandise. Given the need to minimise stock-outs, reduce decision-making times and provide the most efficient assortments, stock-holding levels within an airport environment are prone to gradually increase. A need therefore exists for regular measurement relative to the previous year's figures and current budgetary targets.

Maximising buying power to achieve volume discounts and better prices represents a critical success factor for the buyer. To ensure this is achieved the buyer has to monitor and control both the gross and net margins. Achieving the right assortment in the right place at the right time in the right quantities has a number of indirect costs that can affect performance. Distribution, transport, security and financing costs may all represent significant add-on costs that need to be accounted for in buyer negotiations.

Evaluating Supplier Success

To ensure that the required levels of service are maintained throughout the trading period, a number of retailers attempt to monitor the performance of their suppliers through a Vendor Rating System (VRS). This is a method of determining the performance of each individual supplier against a range of pre-determined criteria. All suppliers, whether they deal directly with the operator or whether they come through an agent, are monitored and evaluated. Typical criteria include the number of deliveries that were inaccurate, the number that were late, the quality of sales support, or the number of returns. The period evaluated varies by operator, however a 12 month review remains the most common. The responsibility for the review falls upon those individuals who have direct responsibility for dealing with suppliers, and typically this would be the warehouse manager or the warehouse supervisor. The management team then evaluates the reviews. Vendors have the opportunity to respond to the comments made. If a problem is identified, then discussions take place and agreement is reached as to what corrective actions need to be taken by the supplier and if necessary by the retailer (Figure 5.3).

	Excellent	Very Good	Good	Fair	Poor
Service					
Sales Support					
Speed of Response					
Information Provision					
Knowledge of Vendor					
Accuracy of Delivery					
Accuracy of Documents					
Shorts / Over					
Repairs / Replacements					

Vendor Rating Form

Please return by post or fax to: Duty Free Manager, Commercial Department

Supplier's Name _____

Comments_____

_____

_____

Buyer _____   Date _____

Supplier's Response _____

_____

_____

Signed _____   Date _____

Figure 5.3   *Example of a vendor rating system.*

Supply Chain Relationships within Airport Retailing – Some Concluding Thoughts

The operation of the supply chain within airport retailing cannot be described as the most proactive nor technologically advanced within the retail sector. While companies such as Heinemann operate centralised warehousing and JIT systems, such practices are not widespread. The growth of passenger traffic volumes has already provided the impetus for change and forced many airport retailers to reassess their approach to supply chain management. In order to remain competitive, airport retailers will need to invest further in information systems and attempt to improve the flow and handling of stock through the supply channel. The areas of most immediate priority include:

- An attempt to better forecast traffic and events in order to match closer the inventory levels with anticipated demand.
- The development of information systems that allow a monitoring of the rate of sales, stock in hand and re-ordering times. This will allow a movement toward minimum stock holding and more accurate deliveries.
- Direct supplier delivery to the shopfloor for a select range of refrigerated and short shelf-life products. This will free warehouse capacity and eliminate investment in chilling equipment.
- A total distribution cost approach to review all physical distribution functions with a view to identifying cost and service benefits.
- A movement by international airport retailers towards a global pricing structure for their major international branded products.

The extent to which these improvements become adopted practice and manifest themselves within the culture of airport retailing remains partially dependent upon the relationship exacted between the retailer and its suppliers. Placing the airport supply chain within the theoretical framework identified previously allows a number of observations to be made. There is little evidence of the vertically integrated corporate channel within European airport ownership. Airport operators have sought to control and administer the supply chain through the development of other forms of relationship. Such an approach reflects the broad, general trends identified within other sectors of European retailing over the past decade. The use of conventional channels appears to be most in evidence within airport retailing. Relationships remain transactional in nature and in the case of tax-free shops involve many small suppliers providing merchandise with few additional services or benefits.

International suppliers therefore derive much of their market power from the high-profile image that their brands convey. There remains little evidence of airport operators using own brands as a competitive weapon; their introduction remains fragmented and very much experimental in nature. However the emergence of multi-national airport operators may encourage the development of own label generic products.

Retailers and airport authorities have focused upon their core business. By centralising the buying they exert power on the basis of the volumes demanded. Smaller suppliers exhibit a greater degree of dependency upon airport operators and retailers, and asymmetrical relationships are very much in evidence. Significant opportunities therefore exist to progress the supply chain beyond the traditional market exchange-based dyad towards a greater form of integrated partnership across the whole of the supply chain.

6 Human Resource Management Issues in Airport Retailing

As I patiently stood in the duty-free queue in Schiphol airport I was aware that the passenger in front of me was Japanese. As it came to her turn to be served, the young Dutch assistant spoke to her in her native language. It transpired that the valued customer had been to Italy prior to arriving in The Netherlands and wanted to pay for half her purchase in lira and use her credit card for the remainder. The assistant took her money, immediately calculated the outstanding amount due and debited her card accordingly. The whole process took less than a couple of minutes. Having wrapped her products for her, he bid her 'Sayonara' before turning to me and saying 'Good Afternoon Sir, How may I help you?'

Introduction

The contribution that good human resource (HR) management can make to the efficient functioning of a business remains a well-developed theme. The importance of selection and recruitment, training and development, morale and motivation are key elements in the guide to better HR practice. The objective of this chapter is to examine these practices from an airport's perspective. As in previous chapters it will be possible to identify similarities between the employment practices of airports and other retail sectors. However, because of the unique nature of airport trading, alternative HR approaches are often required.

As has been emphasised throughout this book, the level and quality of customer service are considered to be critical success factors in airport retail operations. As people remain the key to quality service provision in retailing, it follows that every aspect of the human resource management function must be guided by customer service requirements. In the context of airport retailing, these requirements revolve around the skills and attitudes detailed in the service exchange. The ability to put people at their ease in an unfamiliar environment, a fluency in languages, politeness, friendliness, speed, efficiency, accuracy and a positive attitude are all factors that need to be in evidence. Such attributes must be identified in the retailer's selection and recruitment criteria and developed through induction and on-going training programmes. Work practices and operational procedures must enable staff engaged in the service exchange to use their initiative to ensure full customer satisfaction.

Attempting to develop a general overview of retail employment within airports remains difficult. This is primarily due to the differing labour market structure and legislative provisions that govern different countries. Even within the EU there remain significant differences in labour composition as well as in working practices. The structure of employment has also changed significantly with many countries identifying an increase in the number of women entering the labour force and a growth in the number of part-time positions available. Within Europe, significant differences can be identified in the number of part-time employees working within retailing. Eurostat (1994) highlights the difference between The Netherlands where 47.3% of the retail workforce are in part-time employment and Greece where the percentage is only 3.3%.

With such significant differences in the composition and structure of labour markets it remains legitimate to question whether there is any possibility of developing a generic understanding of retail employment within airports. The diversity of different retail practices combined with different institutional and legislative regulations accentuates this difficulty. Considerable research has consequently been devoted to developing a conceptual framework of the labour market that both accepts and accommodates these differences within its theoretical boundaries.

A significant contribution to achieving this aim has been made through the concept of segmentation. Labour segmentation theory developed as a reaction to the orthodox, neo-classical treatment of human capital. The relative merits and weaknesses of each approach have been discussed elsewhere and are not the focus of this chapter (Becker, 1964; Hunter and Mulvey 1981; Peck, 1989, 1996; Morrison, 1990). What is provided here is an overview of the major theoretical components of contemporary segmentation theory and its applicability to the airport sector.

A Theoretical Understanding of the Labour Market

The starting point for the development of a theoretical understanding begins with the concept of the internal labour market. Originally based upon the work of Slichter (1950), Lester (1952) and Kerr (1954), internal labour markets were seen to be a strategy employed by some companies to reduce the risk of losing key categories of worker through competitive pressures. The objective was to shelter specific groups of employee from the economic demands of the open market by providing them with favourable terms and conditions. Entry into the internal market is defined by the organisation and, once achieved, the pricing, allocation and training of labour are not affected by the market but are governed through a series of institutional rules and procedures. The internal market is distinguished from the external when labour is no longer controlled by economic variables but by the organisation itself.

Gaining access into an internal labour market is through a 'port of entry'. Entry at a specific port provides access to higher positions through an internal promotion system. Piore (1975) argued that the existence of a defined career structure for individuals provides benefits for both employer and employee: internal labour markets provide employers with greater flexibility, while employees are equipped with a range of specific company skills, which makes it difficult for them to transfer to comparable positions elsewhere. Once workers have begun to move up the career ladder, switching to another firm often becomes less attractive as they may be relegated to a lower port of entry. The internal labour market can therefore place a voluntary tie upon the individual by making it unattractive to leave the firm. The practice of restricting entry in this way also allows firms to use their internal markets as screening devices against opportunistic labour. Those workers who were hired in error can either be dismissed or the firm can minimise its losses by halting the progression of an individual on the career ladder (Wachter, 1974; Williamson, 1975).

The notion that specific groups of workers were governed by institutional rather than economic rules remains central to an understanding of segmentation. This theory was further expanded upon by Doeringer and Piore (1971) who maintained that a duality existed within the labour market between different groups of workers of distinct social composition. The most basic distinction was between the *primary* and *secondary* sectors of the labour market. Individuals working under primary conditions not only gained entry into the internal labour market but also received relatively high wages, secure employment and related benefits. Employee turnover in comparison to the secondary sector was low. The secondary sector was characterised as comprising low status and poorly paid jobs that experienced high labour turnover. Such conditions were not attached exclusively to any single industry or organisation and it remained possible to identify both sectors and firms that operated a combination of both secondary and primary employment conditions.

This basic dichotomy was expanded upon by Piore (1975) who maintained that the primary sector could be further segmented into upper and lower tiers of employment. The upper tier is typically composed of those in managerial work or qualified professions; it is distinguished from the lower tier by higher pay, status and promotion opportunities. Upper tier posts are also distinguished from the lower tier by the absence of elaborate sets of work rules and formal administrative procedures. Job demands for primary, lower tier workers place emphasis upon stability and routine. While the conditions of employment are relatively secure, the work itself is often repetitious, rule bound and lacking in interest. Unlike the upper sector, formal educational qualifications are not a fundamental prerequisite for lower tier employment, with performance and experience also contributing to career opportunity.

As with the primary sector, the secondary sector is also structured into tiers. Employment is both hierarchical and segmented, with jobs differing in both quality and quantity. A degree of differentiation therefore exists within the secondary sector. The terms and conditions attached to the job will be dependent upon a variety of internal and external influences, including the availability of labour, the strength of the overall economy and the nature of the organisation itself. What is of importance in this context is the dynamic nature in which labour markets operate. The terms and conditions that surround employment categories are not fixed and change as market conditions change. Thus in periods of labour shortage, employers may increase the attractiveness of the package they offer to employees (Pyke, 1986, 1988).

Michon (1987) maintained that the existence of a secondary labour market provides employers with a series of material benefits. The principal purpose of the secondary sector was to provide flexibility and to allow organisations to cope with the peaks and troughs of the trading calendar. Employees in the secondary sector may be hired by the hour, day, week or month. A variety of different employment contracts exist to provide the employer with the maximum level of flexibility. These may range from a fixed-term period of work to the use of temporary, contracted staff provided through private sector agencies. For example, in the retail sector, part-time employees have been successfully used to cover the busiest trading periods. Paid by the hour, they undertake short, part-time shifts to cover periods of peak customer flow.

In addition to providing a degree of flexibility, secondary sector employment can provide significant savings on labour costs. In the majority of instances the work itself is offered on a part-time basis. Because of the lack of hours worked, part-time employees are often ineligible for sickness, maternity or holiday benefits. Recent EU social policy directives which have become law in a number of member states have attempted to rectify this issue. There has been an attempt to ensure that part-time employees receive the same pro-rata terms and conditions as full-time workers. The attractiveness of secondary workers as a low-cost, flexible labour force is reinforced by their easy substitution. While economic conditions may regulate the overall demand for labour, the low level of skills typically required for secondary employment allows its easy replacement.

The strength of the dualist approach lies in its ability to focus upon the job and not the worker, moreover it highlights the importance of institutional processes in understanding labour market operations. One weakness of the dualist model highlighted by Peck (1996) was its inadequate dealings with supply side factors and the role of the state. The concentration upon production, technical efficiency and employer practices represents a demand side perspective. In addition, an understanding of labour market processes is predicated upon an acceptance that other factors are contingent to its operation. Supply side factors can include the availability of local labour, the influence of trade unions, the level of state benefits as well as the role of the household in influencing labour market participation. Peck (1996) also highlights the role of the state in influencing labour market structures. Labour contracts, industrial relations and employment legislation are among the areas where the state can have a direct impact upon the labour market.

In addition to its hierarchical and segmented nature, the labour market displays a distinct social composition. Women workers are disproportionately represented in the secondary sector (Freathy, 1993; Broadbridge, 1996). A number of studies have attempted to identify the factors behind this allocation process (Wilkinson, 1981; Labour Studies Group, 1985). Such research has revealed the enormous complexity behind labour market allocation. Traditional theoretical analysis viewed the allocation of labour as a wholly demand side process. Employers had control over both the quality and the type of labour they employed. Rubery (1978), however, maintained that while demand side factors played a prominent role in the allocation of labour, supply side influences are also integral to understanding labour market structure. This is not to argue that the relationship between the supply of, and demand for, labour is symmetrical. Typically the demanders of labour have greater control over the market. The relationship is therefore best viewed as an asymmetrical one that is reliant upon a degree of interdependence for its operation (Peck, 1989). For example, demand side issues such as the operation of collective pay agreements and union negotiations can play an important role in structuring the labour market. In addition, however, supply side influences, such as responsibility for social domestic arrangements, mean that part-time employment often lends itself more favourably to female employment (Dex, 1988).

While differentiation based on gender remains the most obvious social segregator, other divides within the labour market include both race and age. Forms of segmentation are therefore not mutually exclusive but mutually reinforcing. Furlong (1990), for example, notes how traits such as physical stamina or family responsibility may also be used to determine an individual's suitability for employment.

The labour market may therefore be viewed as being both hierarchical and segmented. Its structure however is not static and may be viewed as a dynamic operation. The conditions attached to a particular job will vary over time as will the nature of the work itself. Employers and employees seek to ensure a 'fit' within the labour market. Retraining, multi-skilling and increased mobility are all seen as routes to achieving the best match between capital and labour. No single influence is responsible for such change, with both supply and demand side factors structuring the operation of the labour market (Rubery, 1978; Peck, 1988; Broadbridge, 1998).

Contemporary Changes in Retailing: Employment Implications

Peck (1996) contends that segmentation theory is based upon the notion that multi-causal factors are responsible for labour market outcomes. Within the context of the retail sector there have been a number of strategic and operational developments that have had employment-related consequences.

Retailing has become increasingly competitive as companies have expanded beyond their traditional, national boundaries and have sought increasingly to trade in the international arena. The expansion of companies such as Laura Ashley, Next, Body Shop, Aldi, Netto and Lidl highlights the ability of companies to trade successfully in a number of different operating environments. Accompanying this development has been the continued concentration of market power and the dominance of large, multiple-outlet retailers. Sectors such as food and clothing have become increasingly dominated by a limited number of firms.

As Table 6.1 illustrates, the employment implications for these developments are significant. In order to remain competitive there has been an increase in the number of part-time employees working within retailing,

Table 6.1 *Employment implications of retail change.*

Retail change and trend	Employment implication
Strategic change	
Concentration	Market share and power held by a limited number of companies. Employment dominated by large organisations, set standards for the industry
Centralisation	Centralisation of functions leading to the separation of conception from execution and the removal of decision-making responsibility from stores. The creation of specialist functions at head office, such as buyer, merchandiser and space planner
Operational change	
Size of store	Growth of large stores, increased numbers employed on one site
Longer opening hours	Increased use of part-timers, use of specific labour market groups
Service/value competition	Longer working hours leading to increased shift work and part-time employment
Technology	Greater control allows management flexibility. Some instances of deskilling

Source: adapted from Freathy and Sparks (1994).

while at the same time a growing polarisation between management and shopfloor staff has emerged. The existence of a primary/secondary divide within sectors of retailing has previously been identified (Freathy, 1993; Freathy and Sparks, 1994). Retail managers enjoy primary employment conditions and have a defined career path. In addition they may receive a range of benefits as part of their remuneration package (Table 6.2). In contrast, many shopfloor workers occupy a secondary segment of the labour market; working part time in an unskilled position, they have little access to the ports of entry that constitute the internal labour market.

The retail sector also displays a distinct social composition (Broadbridge, 1995, 1996, 1998). Managerial positions based in the primary sector tend to be full time and occupied by males. In contrast, the increasing demand for part-time employment in the secondary sector has been met by utilising a number of labour market groups. Women have found working specific shift patterns fits in with their other household demands. Retailing has also traditionally represented an initial entry into the labour market for young people leaving school.

Table 6.2   *Financial package for general manager in medium/large food superstore, 1996.*

Salary	£60,000 per annum consolidated (no premiums)
Overtime	Double time for Sundays Fixed payment of £100 for bank holidays
Benefits	Employee profit share – average 6% Executive share options Annual bonus (up to 30% of annual salary) Company car (BMW 5 series, Audi-coupe, Volvo 960, Senator); all costs including petrol covered Staff discount card (10% off over £3) 28 days paid holiday per annum Personal Equity Plan Company pension scheme Subsidised staff restaurant 5% discount at travel agents

Source: Company document.

Employment in Airports

The demands upon an employee within an airport retail environment remain significantly different from those in the high street. Not only is the volume of customers significantly greater, but the range of languages, cultures and currencies remains far in excess of what is experienced in the domestic market.

The demands upon an airport retailer also differ from the high street. The requirement for staff to work both night and day shifts at an out-of-town location means that recruitment, training and staff retention measures remain issues of central operational importance. Airports are significant employers of staff, either directly through the services they provide or indirectly through the businesses they attract to the airport. Add to this the multiplier effect upon local suppliers, and the importance of airports to the social and economic welfare of a region becomes apparent. The number of persons employed directly by an airport authority will depend upon the number of services subcontracted to third parties. Airports that have a high number of concessionaires within their terminals tend to employ fewer persons directly. For example, out of a total of 13,000 persons working at Copenhagen airport, only 1230 are employed by Kobenhavns Lufthavne A/S, the airport operator.

For those airport retailers with trading operations abroad, differences between the national and international markets have made the monitoring of activities and the co-ordination of policies an essential management task. While many retailers have made significant progress in this area, many difficulties still remain. Overseas outlets will differ significantly in terms of culture, management style, level of union representation and employment structure. In addition, retailers have to develop a workable balance between local and expatriate staff. They aim to ensure that the organisation remains sensitive to the cultural demands of the country while at the same time imbuing the values of the company to all staff abroad.

Labour represents the second or third largest overhead for the airport retailer, after the cost of the merchandise and the airport concession fee. The control and utilisation of labour therefore become of prime importance in maintaining a competitive advantage. The staffing structure for an outlet will depend upon the airport's operating hours, the number of departing international flights and the trading conditions imposed by the operating authority.

The employment structure at an airport outlet will depend upon a number of factors, including the composition and volume of passengers travelling, the size and layout of the retail unit, the type of products being sold, the IT and EPOS technology used, and the quality and flexibility of staff. Consequently, there remains no single generic structure that can characterise staffing at an airport. The following illustrations are examples of duty/tax-free retail management operations within different sizes of airport. The separation of strategic conception from operational execution, identified in a number of retail sectors, is also in evidence in airport retailing. In each case, the management structures described below are part of a larger retail organisation. Typically these companies will have centralised buying, IT, distribution, business development, financial management and store design function.

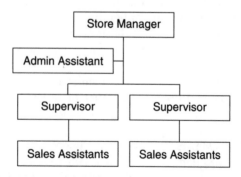

Figure 6.1   *Duty free outlet in an airport with approximately 250,000 departing passengers.*

Figure 6.1 would represent a small regional airport that would have an annual throughput of approximately 250,000 international departing passengers. Typically it would rely upon a limited route network of scheduled flights and a larger programme of non-scheduled, charter flights during the summer, Christmas and winter holiday season. A retail operation in an airport of this size may employ between 14 and 16 people. The store manger would be responsible for all aspects of the running of the store, including sales, merchandising, stock control, shrinkage, local promotions and maintaining pre-agreed budgets.

The administrative assistant would be responsible for all invoicing, cash control and ordering of stock. Given the limited passenger volumes in the airport, it would be unlikely for such an outlet to have a separate bonded warehouse and it may rely upon either direct delivery from a composite third party wholesaler or have a bonded stock area attached to the shop. Supervisors would be responsible for controlling staff and product areas on the shopfloor. In addition, they may be tasked with checking goods in when delivered from bond. Sales assistants would be responsible for check-out operations, customer enquiries and the merchandising of stock on the shopfloor. This latter task includes both the visual presentation of existing products as well as the restocking of shelves. Given the high product demand associated with duty free, this often accounts for a disproportionate amount of labour time.

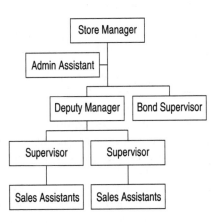

Figure 6.2   *Duty-free outlet in an airport with approximately 1 million departing passengers.*

Figure 6.2 is an example of a duty-free outlet in an airport with approximately 1 million departing passengers. The opening hours would be longer than in Figure 6.1 and the role of the deputy manager would be to cover absences of the store manager. An additional administration assistant may also be employed with responsibility for financial control of the outlet. With passenger volumes approaching one million, a dedicated, bonded warehouse becomes a viable and arguably necessary option. This remains the responsibility of the store manager. The number of night-time departures would be limited to the charter flights in the summer period. Consequently 24 hour opening would not be common and would occur during specific periods of the year. The store may employ between 25 and 30 people, depending upon the time of year.

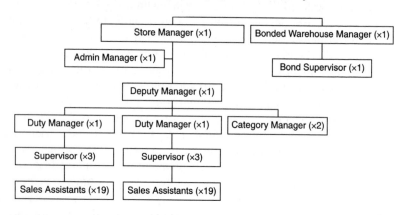

Figure 6.3   *Duty-free outlet in an airport with approximately 3 million departing passengers.*

As the volume of passengers increase then the structure of the labour market also becomes more complex. In Figure 6.3, store managers no longer have direct control of the bonded warehouse as the anticipated volumes sold in an airport of this size require it to be handled by a separate function. In addition, the retailer may have category managers responsible for one or more product areas, such as perfumes/cosmetics/skincare or giftware/souvenirs/soft toys. The role of the category manager is to liaise with both company buyers and external suppliers, ensuring that issues such as merchandising support is provided and that the promotional calendar agreed for that outlet is adhered to. There is also likely to be a larger number of night-time departures and early morning flights as well as increased weekend activity. Shift patterns are therefore an integral part of the labour market operation. With departing passenger volumes of between one and five million, the retailer may employ between 50 and 80 people.

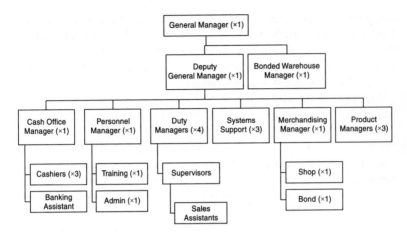

Figure 6.4 *Duty-free outlet in an airport with approximately 7 million departing passengers.*

Figure 6.4 illustrates the employment structure of a duty-free outlet in an airport with approximately 7 million departing passengers. The retail operation will have a defined functional, reporting hierarchy and may expect to employ between 140 and 190 people, depending upon the time of year. An airport of this size commonly has a 24 hour operation and requires employees to work shifts. Because it will typically have a combination of summer and winter charters and year-round scheduled flights, the total numbers employed will vary. In addition to those persons employed by the company, it is also common practice to have a number of personnel provided by suppliers. The fragrance houses in particular make use of this method of direct selling in order to promote their products. While working in the retail outlet, these staff operate under the control of the store's management who determine their working hours and roster duties. While it is their primary aim to promote their company's own products, they would always be expected to deal with any requests from the customer. For an airport with approximately 7 million passengers, the duty/tax-free outlets may have up to 30 supplier representatives working.

In Figure 6.4, the role of the store manager has been removed. The overall responsibility for the store lies with the General Manager, although there is only limited operational control. The function of a General Manager in this size of airport is primarily to manage the overall business at the airport, and to liaise with the airport authorities, public bodies and retailer's head office. The day-to-day operational running of the store is the responsibility of the Deputy General Manager who will have four or more duty managers reporting directly. This scale of operation may merit an on-site Personnel and Training function. Its responsibility is to recruit and select sales staff, and provide company induction and on-going sales and product training.

The volume of sales in this size of airport also allows for a separate cash office and banking function, with a Cash Office Manager being responsible for the financial management of the outlet's revenue. Systems Support is responsible for all technical aspects of a store. In smaller outlets this function may be combined with an administrative role. It is their responsibility to maintain all point of sale equipment and supporting technologies such as price look up (PLU) and e-mail facilities. An outlet that serves over 7 million departing passengers may also have a separate merchandising team that is responsible for maintaining the general decor and layout of the store. The group may be responsible for signage, promotional material and store atmospherics. In addition it may be tasked with dressing all windows and show-cases, producing all ticketing and mounting product displays.

As the volume of departing passengers further increases, the staffing structure within a duty/tax-free store will become a further variation of Figure 6.4. For example, a store may employ two Deputy General Managers and six Duty Managers. The Systems Support team may merit its own manager and there may be a separate Administration function.

Staff Numbers

Because of the long trading hours that the majority of airport retail outlets are required to operate, the total number of staff employed in a store tends to be greater than in a domestic unit of equivalent size. Many airports have the requirement that their retailers must be open for all international departing flights. Such a policy can cause significant staffing difficulties when flights are delayed. In airports with the greatest passenger volumes and the longest trading days, it is not uncommon to have 60–70% more staff employed than in an equivalent high street site. The number of persons employed by a retailer will be further influenced by the location of the stores within the terminal building. If the retail facilities are located centrally then the concessionaire is often able to enjoy scale economies in the allocation of labour. If however they are located throughout the terminal and at the departure piers, then there will be a requirement to employ proportionately more staff.

Determining the total number of persons required to staff an airport retail outlet is based upon a number of factors. The first is the total level of sales that passenger flows are expected to generate. Wage budgets will be set as a percentage of this figure. Although the amount will vary between retailers, typical employee costs remain between 4% and 8% of turnover. Managers will be expected to recruit staff within these set financial parameters. A second factor influencing staffing levels will be the level of customer service that the airport authority expects the retailer to provide. The attempt by many airports to provide better quality merchandise in a better quality shopping environment has meant that value delivery has become an integral part of their trading strategy. Tender submissions may therefore require the bidder to guarantee certain minimum levels of service, for example a maximum queue length and the ability to open all check-outs if required. The airport may also demand that certain products such as perfumes be sold over the counter rather than on a self-service basis. While supplier representatives may assist in product demonstrations and selling, such a requirement will also influence the total number of people employed in store and consequently the wage budget. Other factors influencing staffing levels will include the timing of the flights, the peaks and troughs of the airport's traffic patterns and the layout of the store itself.

Airport retailers face a number of unique challenges in attempting to maintain wage budgets. In particular the problems of flight delays present a number of difficulties. As noted previously, short delays (e.g. up to an hour) can have a beneficial effect upon retail sales within an airport. Passenger spend is likely to increase as customers purchase additional products such as confectionery and beverages. Longer delays, however, have a negative impact as frustration levels rise and retail sales consequently decline. For those airports not open on a 24 hours basis there is the proviso in many concessionaire contracts that retailers have to continue trading until the last flight has departed. This means that employees in these airports may have to work additional hours, often at premium rates. The option of extending part-timers' hours to cover such events (thus avoiding premium rates) is often not open to many airport retailers as the majority of staff are full time.

An additional problem for many employers is that of delayed flights arriving towards the end of a shift. The EU working time directive requires that employees on shift working must be given an eleven hour break between shifts. Flight delays, usually occurring late at night, can extend a shift significantly and leave insufficient time before the start of the morning shift to allow compliance with the working time directive.

A further difficulty for retailers is the ability to accurately gauge customer flow on an hourly, daily and weekly basis. While the total number of passengers travelling through an airport can be calculated and segmentation studies have allowed detailed consumer profiling, the extent to which this assists labour scheduling on a day-to-day basis is limited. For example, while the retailer will know the times of flight departures and the capacity of aircraft, information on the level of occupancy or the passenger manifesto will not be available to either the retailer or the airport authority.

Staff Contracts and Working Hours

There are broadly three categories of employee within airport retailing:

- *Permanent employees, working full time or part time*: these represent the overwhelming majority of employees within airport retailing.
- *Contract employees*: usually on a renewable fixed-term contract for a specific period, e.g. 6–12 months. Typically these staff are used to fill long-term vacancies such as maternity leave.
- *Casual employees*: these are recruited depending upon the needs of the business. Management, using forecast data, will assess the level of staffing required and call persons in as necessary, often at very short notice. Because of the issues of security clearance in some airports, casual workers have a higher propensity to be employed landside.

One area where the staffing structures of an airport and a domestic retailer are noticeably different is in the employment of part-time workers. While part-time employment has increased in significance in many European countries (Eurostat, 1994; Dawson, 1995; EIRR, 1997), its role within the airport environment remains limited. The main advantage of employing part-timers is the flexibility they provide the retailer in dealing with the peaks and troughs of the working day. However, in larger, international airports especially, there remain few periods when demand would justify employing part-timers. Consequently, full-time retail employees may constitute 70–80% of the airport retailer's workforce. Part-time employment is concentrated at specific times of the year, most notably during the peak summer months and the winter holiday periods when traffic flows are at their greatest.

The threat of intra-EU duty-free sales being abolished in June 1999 has begun to alter airport labour market trends. A fear among many airport retailers is that if duty free is abolished there exists no comparable product able to generate similar sales volumes and revenues. While still employing full-time staff, there is a growing reluctance on the part of some retailers to offer individuals a permanent contract. As a way of maintaining future flexibility and ensuring that the company does not have a surplus of labour, the use of fixed-term contracts has become increasingly popular in those countries in which it is permissible by law (EIRR, 1997).

The use of subcontracted, agency labour is not widespread within airport retailing. Despite the potential flexibility it can provide, retailers have opted to employ their own staff. The reasons for this include cost, the lack of control over labour quality as well as the regulation requiring all personnel to be security checked before working airside. The complexities surrounding the maintenance of a pool of security-vetted agency staff have led the majority of retailers to avoid subcontracted labour.

The shift pattern for employees will vary both by company and by size of airport. For example, a small regional airport with irregular flights may not be able to specify exact shifts. In this instance the retailer, who will know flight schedules in advance, may be obliged to guarantee a fixed number of hours per week and adapt labour scheduling to fit in with flight departures. In larger airports a retailer may operate a variety of shift patterns. For example, five, seven and a half hour shifts with the following two days free remains a common option, while others may prefer four shifts of nine and three quarter hours with three days free. A third option practised by retailers is three eleven hour days with the following three days free. The choice of shift is therefore dictated by the operating hours of the airport, its flight pattern, the demands of the retailer and the availability of labour. A three shift, weekly operation for a tax-free retailer is illustrated in Figure 6.5. The structure allows a one-hour overlap between shifts. This overlap may be used in a variety of ways, most typically to allow staff to complete cash declarations and reconciliations, to discuss stock replenishment, for training or to provide staff with briefings and information sessions. One of the advantages of shift working in an airport is that it allows employees at least two full free days off in every seven, compared with the more common situation where time off is divided through the working week.

Maintaining staffing levels and maximising productivity represent a challenge for airport retailers. Staff retention is normally highest when employees work regular hours and are able to predict their shift patterns well in advance. Because of the nature of the industry such guarantees can rarely be given. This frequently leads to staff absenteeism and turnover being higher than in other areas of retailing. The problems that this

	0600	0900	1400	1500	1800	2100	2300
Monday	◄———	Shift 1	———►	Shift 2			———►
Tuesday	◄———	Shift1	———►	Shift 2			———►
Wednesday	◄———	Shift 3	———►	Shift 1			———►
Thursday	◄———	Shift 3	———►	Shift 1			———►
Friday	◄———	Shift 2	———►	Shift 3			———►
Saturday	◄———	Shift 2	———►	Shift 3			———►
Sunday	◄———	Shift 1	———►	Shift 2			———►

Figure 6.5 _Shift pattern working for a tax-free store in an international airport._

creates are twofold. First, difficulties are often experienced in getting persons to apply for positions within the airport. The rates of pay offered by airport retailers are broadly comparable with other areas of retail, yet the hours are often unsociable and the location involves more travel and expense. It may therefore be necessary for retailers to consider alternative approaches to external recruitment. Strategies may include focusing upon non-traditional labour market groups such as older workers, targeting staff in other organisations, advertising in specific residential areas, improving employment terms and conditions, and offering assistance with child care and relocation.

Secondly, once staff have been recruited, getting them to remain in the longer term becomes an issue. To arrive for a 0600 start in winter at a location that may be some distance from home has little appeal for sales staff. In addition, many staff are reluctant to work late and travel home unaccompanied. Where the airport has good public transport links then this situation has been partly ameliorated, while in other instances, retailers may offer a shuttle bus or taxi service for those employees working very late or on early shifts.

To help attract and retain staff, many retailers are using a range of human resources management strategies including job-sharing, personalised hours, additional leave and holiday privileges (e.g. paternity, emergency, eldercare and compassionate leave), employment breaks, sabbaticals and secondments. Other initiatives include flexible working hours, child support facilities, parenting support, credit union facilities, health care, family events, discount schemes, assistance with loans/college fees and work experience for children of staff. The benefits of these practices are wide ranging and include improved morale and productivity, reduced employee stress levels, a decline in sick-leave and absenteeism, and increased employee flexibility.

Retailers have also attempted to combat absenteeism by structuring their operations into small working groups. One problem with larger duty-free outlets is their impersonal nature. Staff may not know their colleagues as each individual may work on a range of different shifts. By developing groups of around 20–30 persons and keeping these employees together, retailers have been able to foster closer working relationships between individuals. These small teams function along the same lines as quality circles and help to identify operational problems, areas for improved customer service, and ideas for increased efficiencies and cost savings. Coherence has been built upon by creating competition between the different groups. Sales results may be published and prizes may be offered to the most successful groups. While team building schemes have been successful in improving staff motivation through greater empowerment and involvement, care must be taken to ensure that store managers and supervisors fully understand the concept and are properly trained in its implementation in order to avoid any loss of management confidence or morale.

One airport retailer in the UK identified that there was a direct correlation between absenteeism and the eligibility of staff to receive pay through the company's sick leave scheme. It also discovered that the level of staff absenteeism was directly proportional to the seriousness with which it was treated by the store's duty managers. For example, if return-to-work interviews were conducted with staff after every absence, then the proportion of staff absent from work reduced significantly.

Pay and Salary

The majority of pay for employees comprises basic pay, shift allowances and sales commission. While there are examples of retailers who pay a single flat rate to staff, these remain in the minority. How the three remuneration variables are weighted depends upon the individual retailer. There is only limited evidence of retailers offering the types of incentive packages to store managers similar to those found in other retail sectors (see Table 6.2). Managers in the largest airports may have entitlement to a company car, health insurance, a non-contributory pension scheme and salary bonuses. In some instances this may be up to 30% of salary if profit targets are met. While these incentives normally apply only to managerial staff, some retailers offer a flat rate bonus to all staff, based on the performance of the company as a whole and the individual store's out-turn relative to target. Staff employed at an airport may also be entitled to favourable terms from many of the commercial concessions at the airports, such as free banking services, commission-free foreign exchange transactions and discounted shopping.

Because of the high passenger throughput in an airport, sales volumes remain disproportionate to store size. For example, one electrical retailer estimated that it sold between 18 and 20 times more 'walkmans' in the airport than in its equivalent high street outlets. Because of these large sales volumes, the commission rates for staff will be calculated differently from those in the domestic market. One retailer with stores in both the airport and the high street had benchmarked the sales commission for its airport shops against the largest superstores in the company's portfolio. An employee working in a 185 square metre unit in an airport would generate sales equivalent to a 2800 square metre out-of-town unit.

There remains no single method of dividing commission between staff, each approach having its own advantages and disadvantages. Some retailers view the salesforce as a team, so all commission earned is pooled and divided equally at the end of the trading month. While appearing to be equitable in its approach, this method has been criticised on its inability to differentiate between the best and worst sales assistants, and may represent a disincentive to some staff. An alternative approach has been to allow individuals to earn commission on their own personal sales. While this overcomes the problems highlighted above, it also creates a number of difficulties. First, customers may consult with a number of sales staff prior to purchasing the product, and this may lead to tension between staff as to who should receive the sales commission.

Secondly, the opportunities for some employees to earn commission is greater than others. For example, most duty-free outlets are laid out on a self-service basis with customers taking their purchases to the pay point. However some check-outs attract more customers than others and staff quickly get to know the best registers to be assigned to. Some airport retailers find it necessary to roster staff by register in order to allow all employees an equal opportunity to earn commission. Equally, the sales volumes of products such as liquor and tobacco in most airports exceed those of perfume and other tax-free merchandise. However, the selling effort required from staff can be greater in the perfume and tax-free areas. As a result, many retailers operate a differential commission scheme with lower rates of commission in the self-service areas.

A further source of tension centres on the sales made by supplier representatives. As these staff normally cannot earn commission directly in the store, the retailer's own staff may resent any sales made by these individuals. Some employees have taken a more pragmatic approach and have developed a working relationship with the representatives who allow them to make the sale and earn the commission. One solution attempted by a number of retailers has been to pool all the commission earned by supplier representatives for subsequent division among the retailer's own staff.

Selection and Recruitment

Because of the high work pressure and unsociable working hours, staff turnover within an airport is often higher than in other retail sectors. While staff are recruited through a number of media, the most common remains local and national newspapers. Skills and qualifications for sales staff will differ significantly between company. Some retailers will require no prerequisite qualifications, others will demand the same level of skills as their domestic stores, while some place higher demands than would be expected elsewhere. In The Netherlands, for example, the perfume and cosmetics concessionaire Kappe Schiphol BV has a requirement that all applicants speak a minimum of two languages (with three preferred).

Managers are recruited from a wide variety of sources. The recruitment policies of airport retailers obviously vary but a common criterion has been the ability of individuals to cope with high levels of trading activity. Consequently, many expressed a preference for managers with previous experience in a high-turnover retail environment. For example, retailers in the UK duty-free industry looked to recruit experienced managers from companies such as Marks and Spencer, Sainsbury and Tesco.

The level of activity, especially on the salesfloor, is often both physically demanding and highly stressful. Finding suitably qualified sales staff can therefore be problematic. The way in which retailers have sought to respond to this difficulty has varied. Some have accepted the fact that working in an airport creates high levels of staff turnover. Consequently they have recruited whatever staff they can from the local environs. This form of imprecise matching of employee supply to employer demand has had implications for the delivery of value and the levels of customer service offered. A more proactive strategy has been to identify those individuals most suited to working in an airport environment and assist them in overcoming the difficulties associated with working there (travel, car parking, unsocial hours etc.).

The criteria upon which staff are selected will vary but may typically include:

- level of education;
- relevant experience;
- good communications skills;
- literacy and numeracy;
- personality and confidence;
- customer service awareness;
- ability to work in team environment;
- personal neatness and tidiness.

Staff are usually recruited through national and local press advertising. Some retailers use personnel agencies as well as advertising to existing employees who may have family and friends willing to work in the airport. Where a new concessionaire takes over an existing outlet, the incumbent staff will be encouraged to apply for positions in the new store. Employers are required under the EU directive on acquired rights to offer existing staff their previous positions. The selection methods used by retailers will vary, but can include a CV assessment, aptitude tests, one or more interviews and the checking of references.

Training and Promotion

Deciding upon the skills required to effectively manage an airport outlet has remained the subject of some controversy. One school of thought has maintained that retail products within airports effectively sell themselves. The level of staff–customer interaction (and consequently the level of staff training required) remains minimal. Among some retailers of duty-free products especially, this view has been quite widely held. Because tobacco and alcohol products have displayed high levels of planned purchase, there has been little requirement for staff to demonstrate in-depth product knowledge. More recently this situation has changed, prompted by three significant factors.

First, the sales volumes and rents now expected by an airport authority have meant that many retailers have had to take a more aggressive, proactive approach to selling product. Secondly, a limited number of airport authorities now allow retailers to sell competing products within the same terminal building. While there remains the obvious danger of providing the traveller with too great a choice, given the time constraints under which they operate, it represents an attempt by the airport authorities to stimulate sales through competitive pressures. The responsibility for the retailer is to ensure that its staff is trained to maximise sales by providing the best levels of customer service. The third factor relates to the wider product choice that consumers now have. The expansion of the retail offer within an airport means that consumers may choose to purchase alternative products or services. Not only has this intensified the pressure upon traditional duty/tax-free retailers, but also the sale of high-value products such as watches, fashion clothing, gourmet foods, fine wines, diamonds and jewellery has further increased the demand for knowledgeable, trained sales personnel.

The types of training received by retail staff will vary. Some airport operators insist that all individuals working within an airport undertake specific fire, health and safety training. If a retailer is part of a national chain with domestic operations, then staff at the airport outlet may follow a generic company wide training schedule. Typically, an induction programme for new non-managerial staff will last between one and a half and five days, and may include:

- airport tour;
- store familiarisation;
- customer care;
- financial skills training, including currency and traveller cheque handling;
- EPOS awareness;
- understanding vendor control;
- fire training and emergency drills;
- the organisational structure of the company;
- company history and values.

A programme of on-going training is also practised by some of the more progressive retailers. Typically a sales assistant will follow a task-based study programme and will be periodically assessed by the line manager. This formalisation of training against a set of pre-defined standards has been seen as a useful method of motivating individual workers as well as providing detailed feedback on an individual's potential for promotion to more senior and supervisory roles.

Typically, continuation training would involve:

- customer care;
- selling skills;
- product knowledge;
- merchandising;
- presentation and display of products;
- cash control;
- budgeting.

The priority ascended to assessing the suitability of managers for promotion has also increased among airport retailers. Managerial appraisal schemes have become widely adopted, as has the identification of the criteria against which an individual may be assessed. There remain a number of core competencies against which managers may be assessed during their appraisal. For a manager in a duty-free outlet, these may include:

- change management;
- communication;
- strategic thinking;
- team working;
- functional ability;
- business awareness;
- results orientation;
- leadership;
- personal effectiveness.

Assessment centres have increasingly been used by airport retailers as a means of identifying a manager's suitability for promotion. Building upon appraisal details, an individual's development can be monitored and further examined. Weaknesses can be identified and any skills gap can be rectified through further training.

As Peck (1996) contended, the structure of the retail labour market remains the outcome of supply, demand and institutional factors. Its arrangement is the result of multi-causal influences that differ by company, by airport and by country. The opportunity to draw general conclusions on the nature of employment therefore remains limited.

One of the most significant differences between employment in airports and other retail areas is the method by which employment planning is undertaken. In the UK, for example, retailers have sought to maintain maximum flexibility by identifying the peaks and troughs of the working day and employing part-time workers to meet this demand. The use of part-time labour, common in many European countries, was not part of the employment strategy of many airport retailers. The constant product demand experienced in many airports combined with the requirement for staff to work unsocial hours, not only limits the retailer's requirement for part-time employees but also its ability to recruit enough suitable staff.

The importance of institutional factors in influencing the structure of the retail labour market is also highlighted by the threatened abolition of duty free in 1999. Companies, while still offering full-time employment, remain unwilling to translate these posts into permanent positions until the position has been clarified. In this way, airport retailers have sought to maintain a degree of flexibility in their operations.

7 Future Developments in Airport Retailing

Introduction

One theme of this book has been the attempt to illustrate the relationship between airport retailing, the macro environment and the wider demands of the air transport industry. While retailing continues to remain a response to the cultural demands of the market it serves, the increasingly international nature of the passenger market has meant that airports require a flexible and proactive approach. Constant change is a characteristic of the industry, and airport retailing, subsumed within this broad definition, has been no less affected by these developments. An understanding of the factors that will continue to influence airport retailing into the next century is as much reliant upon macro-environmental factors as sector-specific pressures.

Future Influences upon the Air Transport Industry

Perhaps one of the most important factors from the perspective of the airport operator will be the continued growth in passenger numbers. In 1993 scheduled traffic in Europe accounted for 54% of the total world traffic, by 2010 this figure will still be in excess of 50%. The year-on-year forecasted growth in the total numbers of people travelling by air will average 5.1% between 1993 and 2010, and IATA expects scheduled passenger traffic within Europe to increase by 230%. In addition, the proportion of higher margin business passengers is also likely to increase.

The increasing volume of passenger traffic has a number of implications for the world's airports. The problem of air congestion in particular will intensify. In Asian countries already there are indications that this has become a serious problem. The situation has not been helped by the announcement of Boeing/McDonnell Douglas to halt development on their large 600+ seat supercarrier on the basis that it was unable to guarantee the 78% passenger load factor necessary to make the programme financially viable.

The increases in passenger traffic predicted over the next ten years has also raised a number of safety issues. Concern has been expressed in particular at the lack of co-ordinated air traffic control between emerging countries keen to develop tourism. In particular, the absence of an adequate navigational infrastructure and the lack of English-speaking air traffic controllers represent a threat for future air safety.

To adequately cope with the increased demand for flights and to avoid the congestion noted by Doganis (1993), there will be a need for larger and more efficient airports. This is most likely to be achieved through the continued development of a modern infrastructure and an integrated transport system. In particular, inter-modality is considered to be a prerequisite by the industry, with the airport becoming the nexus for road, rail and air links. By channelling significant funds into its Trans European Networks (TENS) programme, the European Union is speeding the development of these inter-modal networks throughout Europe. One effect of this process will be the switching of travellers to rail and the demise of many of the short-haul feeder air routes. Airports therefore face a dilemma, and the availability of high-speed rail links will be a necessary requirement for those airports wishing to become mainport hubs. However, Bipe (1997) maintains that the impact of rail on the air transport sector could lead to intra-EU air passenger traffic being reduced by over 3% per annum.

The current international passenger hubs of London Heathrow, Amsterdam Schiphol, Frankfurt and Paris Charles de Gaulle are most likely to dominate the air transport market, as they increase their market share at the expense of the regional airports. What this may indicate is a growing polarisation in the future structure of European airports as the largest hubs further dominate the industry. For example, of the 534 airports in Europe, the 16 largest offer as many aircraft seats as the remaining 518 put together.

The response from other airports has been twofold. Some larger regional airports, such as Dusseldorf and Manchester, have positioned themselves as secondary hubs, focusing upon specific groups such as the holiday/leisure market. The method of servicing this market is changing as tour operators make increasing use of scheduled services as well as charter flights. This development has helped these airports to secure long-haul carriers and establish new routes to the Americas and the Far East. As many of these new routes are on a scheduled basis for most, if not all, of the year, this helps to guarantee income during the shoulder months and during off-peak periods. These airports have attempted to compete for traffic by offering competitive aeronautical charges to the airlines, preferential facilities and quick turnaround times.

A number of smaller regional airports have actively sought to avoid competing with the larger hubs. One tendency in the 1980s was for some regional airports to position themselves as scaled-down versions of their international counterparts. In many instances this was unsuccessful, with the airport being unable to gain the same economies of scale, attract similar passenger volumes or gain equivalent levels of investment. Because they lacked a critical mass of passengers they found it difficult to attract airlines, hotels and car hire firms. The experience of Southend airport in the UK underlined the difficulties of a regional airport. Hit by recessionary pressures and the dominance of the larger London hubs, the airport experienced a drop in passenger numbers from 175,000 in 1989 to just over 5500 in 1994. Similarly it saw freight operations drop from 26,000 tonnes carried to just over 4000 tonnes during the same period.

How smaller regional airports react to such pressures will depend upon a number of factors. Not least will be the level of local authority or state aid offered and the extent to which Government is willing to intervene in route scheduling. In the case of Southend, the airport took a commercial decision to redefine its business strategy so that it no longer competed head on with the larger airports. It achieved this by focusing upon the needs of the local business community and attracting support services to locate on the airport site. The development of business parks will therefore continue to represent a central element in the long-term strategy of many airports. Not only do they provide a stable income stream from property leasing but also remain essential for generating air traffic at the regional level.

One other option for the smaller regional airport may be to specialise in specific routes. For example, Eastleigh airport in the UK has specialised in routes to and from the Channel Islands. This specialisation will be facilitated by the process of air transport liberalisation that has been underway in the European Union for a number of years. The final stage in the liberalisation process was completed in April 1997. The objective of this initiative is to further stimulate competition by removing the remaining barriers to market entry. Such a strategy has already encouraged new carriers to enter the market and compete with the established airlines.

The liberalisation process has resulted in the emergence of an expanding, low-cost scheduled airline sector that is expected to grow for at least the next decade. Many of these new market entrants are offering fares which are typically 60–85% lower than the cheapest unrestricted fares offered by conventional airlines. As a result, these new airlines are creating new point-to-point or city-to-city routes using secondary airports situated close to the major cities. The additional number of EU passengers generated from the low-cost sector is estimated to be in excess of 32 million over the next five years (Symons, Travers, Morgan, 1997). It is expected that, so long as these airlines can maintain their large fare price differential with the major flag carriers, they will continue to be successful. The only real limits on growth will be the high start-up costs within the industry, the scarcity of aircraft slots at the major airports and a shortage of reasonably priced aircraft.

For Europe's regional airports, the liberalisation process provides a major opportunity to increase their market share and to gain further specialist routes by attracting the discount operator. While it remains speculative to gauge the reaction from the bigger carriers to this development, the costs of competing directly for such routes would remain prohibitive, especially given the process of rationalisation that has already taken place. Early indications therefore suggest a more pragmatic response, with franchising agreements being concluded between major airlines such as British Airways and Sabena and the discount carriers. The objective is to operate routes on behalf of the major airlines but on a significantly lower cost base.

A greater liberalisation of air transport and an open skies policy has helped stimulate competition among the airlines and reduce the reliance of many airports upon a single large carrier. However, at the same time it has demanded an increasingly proactive strategy from airport authorities as the competition for airline revenue intensifies. As new and existing entrants successfully develop routes and niche markets, the overall effect should be a general lowering of fares as airlines aggressively market themselves to the traveller. Open skies agreements between the US, the EU and its member states should also mean a further proliferation in the number of operators on specific routes.

The liberalisation process will stimulate significant additional growth within the EU and in the potentially untapped markets of Eastern Europe and the Far East. The imperatives attached to the maintenance and running of airports will continue to represent a significant financial burden upon the state and further stimulate the need for private sector involvement. It is estimated that, over the four years leading up to the next millennium, Europe's airports will be required to invest over $20 billion in additional terminal capacity (ACI Europe, 1996; O'Toole, 1997). This requirement is occurring at a time when Governments are committed to reducing public expenditure and is a further motivation towards privatisation.

The air industry is likely to continue to be characterised by both a strengthening of alliances and a consolidation of its main players. Alliances exist primarily between non-competing airlines who agree to develop closer working relationships and to code-share on specific routes. There are increasing instances of code-sharing arrangements involving three or more carriers. For a number of smaller airlines, such alliances have been a prerequisite for their continued survival. Alliances are considered to increase market share while helping to improve capacity rationalisation, lower prices and control costs. Examples of airline alliances are illustrated in Table 7.1.

Table 7.1 *Examples of airline alliances.*

Lufthansa / SAS / United Airlines / Thai / Air Canada
KLM / Northwest Airlines / Air UK / Air Exel
BA / Qantas / America West / Brymon Airways

Retail Changes in the Air Transport Industry

For the consumer, shopping at the airport, whether on departure or in transit, has become an integral part of the travel experience. The year-round daily opening, long trading hours and wide choice all provide many travellers with the opportunity to go pleasure shopping. Passengers travelling for leisure have the greatest propensity to impulse purchase at the airport. Mintel (1997) identifies 'transumers' – consumers who, because they are travelling, think, act and shop in a different way from consumers in the high street. As passengers become more experienced airport shoppers, their expectations will grow, leading to demands for greater retail choice, better quality and service, and value for money. The main concern for airport operators is how to continue to meet these increased expectations in the face of growing competition and continued structural change within the industry.

One of the primary concerns for retailers is the impact that competitive, consumer and structural change will have upon their pattern of trading. For example, while a limited increase in congestion at airports can stimulate sales, the scale of delays forecast by some members of the industry would lead to a negative impact upon retail purchases. Having disproportionately large queues for passport control and immigration will adversely affect an individual's propensity to purchase. This issue has been recognised by the industry and was used to support the argument for the development of a second terminal at Tokyo's Narita airport. Airports like Kai Tak in Hong Kong have acknowledged that the levels of passenger congestion within the terminal work against their ability to generate commercial revenues. The Civil Aviation Department of Hong Kong maintains that the airport could double its commercial revenue ($400 million in 1996) if it had more space available.

Similarly, the implementation of the Schengen arrangements in participating member states has meant the removal of passport and immigration controls on travel within the majority of EU countries. For airports this has led to the physical separation of Schengen and non-Schengen travellers, with a consequent loss of capacity and the unnecessary duplication of facilities. The Schengen agreement, to date, has not worked in the interest of airport retailers, who in order to reach the maximum number of passengers have had to open stores in both the Schengen and non-Schengen areas. Sales to date have been disappointing, with considerably reduced penetration levels.

With an increasing number of airports being privatised, the need to satisfy shareholder and investor expectations will ensure that emphasis continues to be placed upon profit-maximising strategies. Such an approach will have a number of retail implications. Any involvement of the private sector will be based upon the expectation of a reasonable financial return in the short term and the enhancement of the asset value of the airport in the medium to long term. Full commercial criteria will apply to every aspect of development and it is expected that in this changed environment, attitudes towards retailing and other revenue-earning activities will become more positive. Privatisation is also likely to lead to increased regulation of aeronautical and related charges using state-appointed, independent regulators or review agencies. In this situation, the non-regulated activities of airports will become even more central to the airport's financial strategy.

Joint ventures are likely to continue as a mainstream strategy and will grow in importance. Both airport authorities and airport retailers will continue to look for trading partners to co-operate in the areas of purchasing, operating the full retail function or in some instances taking responsibility for the running of the whole airport. Companies such as Saresco in France, Heinemann in Germany, BAA in the UK and Aer Rianta in Ireland are all actively looking for partners worldwide. In many instances the objectives will be to tie up with organisations from the host country. Such a strategy reduces the risks associated with competing in foreign markets, remains politically expedient, as well as being a means of gaining local expertise and market knowledge. The growth of the

Asia–Pacific region in particular provides a number of expansion opportunities. For example, the privatisation of the Australian airports attracted interest from 26 potential bidders, including Aer Rianta, BAA, Manchester Airport, Schiphol Airport and National Express.

For some companies, growth will be achieved through acquisition and merger. Swissair's purchase of Allders' duty/tax-free retailing business for £160 million, the sale of DFS to LVMH for $2.47 billion and the 30% stake of Alpha bought by the chairman of Harrods are indicative of an industry where market concentration is likely to be a continuing trend. Looking to the future, it remains likely that duty/tax-free retailing and possibly the management of the airports themselves will be controlled by a limited number of more powerful organisations who have created market share through a policy of rapid and aggressive expansion.

The airlines themselves will come to represent an even greater threat to the airport authorities and retailers. In their attempt to improve the process of air travel and create a more-efficient 'seamless' and customer-friendly service, airlines have increased the number of services available to the departing passenger. Automatic ticketing and self check-in facilities, for example, have further reduced the time a passenger has to spend at the airport before departure.

More fundamentally, the liberalisation of air transportation and its consequent impact upon air fares, cost structures and route networks will lead to increased price sensitivity and greater competitiveness within the industry. Carriers will find themselves operating on even tighter margins, which will increase their already high dependence upon in-flight duty/tax-free sales revenues. The likely strategy will be to expand on-board sales by adopting a more competitive marketing stance against the airports and aggressively pursuing an increased share of passenger spend.

The proposal to abolish EU duty/tax-free sales, if implemented, will bring the airlines into even greater competition with the airports. Primarily, this will occur in two ways. First, there will be an indirect impact as charter operators include more non-EU holiday destinations in their catalogues. As airlines will continue to sell in-flight duty/tax-free goods on routes outside the EU, operators will choose these locations in preference to destinations within the member states. Secondly, the non-commercial revenues of those airports serving the principal EU holiday destinations will come under even further pressure both from the renegotiation of aeronautical charges as well as the possibility of reduced passenger numbers. For example, airports such as Ciampino (Rome) and Beauvais (Paris) rely heavily upon intra-EU, non-scheduled flights

(66.8% and 89.4% respectively); any destination switching by airlines would disproportionately affect both the commercial and aeronautical revenues of such airports.

The liberalisation of air travel will have an impact upon the marketing strategy of many airports. Despite arguments to the contrary, passengers are not a captive market, a fact illustrated by the low sales penetration rates of many airports. Both retailers and the airport authorities therefore have to continue to be proactive in their approach to the exchange process. Liberalisation has introduced new consumer segments to air travel. The emergence of alternative consumer groups will present airports with the challenge of converting these passengers to airport retail customers. The objective for the airports in the long term will be to maintain their share of passenger spend in the face of more sophisticated and aggressive marketing strategies by the airlines.

This demand will become an even greater necessity as airlines avail themselves of the many new technologies emerging on to the market. Some such as Premiair, the Scandinavian airline, adequately illustrate this; it offers a duty/tax-free ordering service to passengers prior to leaving their home. Premiair customers can order and pay for duty/tax-free goods from a catalogue in the comfort of their homes with the knowledge that the goods will be on their pre-assigned seat on the aircraft when they board.

Airlines remain well suited to this strategy as they have the advantage of having initial contact with passengers through the travel agent network and the ticketing process. The information management technology available through the airline computer reservation system (CRS), coupled with modern direct marketing techniques, provide airlines with powerful tools with which to individualise the contact with their passengers. While not in wide-scale use by the airlines currently, such a strategy has the potential to effectively leap-frog the retail offer within airports and challenge the current dominance of airport retailing. A significant marketing advantage would therefore seem to lie with those providers who can present the retail offer to the passenger first.

Frequent flyer schemes offer airlines almost unparalleled access to business passengers. Many airlines are already using newsletters and contact magazines associated with these programmes to cross-sell holiday and short break travel packages. The potential exists to expand these sales activities to include shopping offers for on-board collection or even home delivery. Airline loyalty programmes also provide access to special or exclusive waiting lounges for their regular members. These areas offer quiet surroundings away from the terminal and in many instances provide free refreshments and full business centre facilities. Virgin Airlines even provides a full range of video and electronic games, plus relaxation services including massage and bath facilities for clients.

Such initiatives are not without their cost in retail terms. The ability to reduce dwell times even further within airports will have a detrimental effect upon retail sales. As has been illustrated within this book, time remains the single most important determinant of an individual's propensity to purchase within an airport. Reducing the time that passengers are in the terminal building by increasing the availability of airline lounges for both loyalty club members and for those travelling on business class tickets will have a negative impact upon sales. Moreover, the opportunity exists to use these lounges to provide an alternative and competing retail offer, thereby further eroding the airport's share of passenger spend.

Computer and tele-communication technologies may offer further opportunities for airlines to increase their share of the retail market through the development of new on-board sales techniques. Already airlines have introduced direct mail catalogues. These allow a wide range of goods to be ordered by passengers while on the aircraft, on arrival at their destination or later from home. The installation on many aircraft of individual interactive video screens also offers airlines an electronic selling medium similar to the home shopping channels on TV. The personal telephone facility available on many aircraft uses credit card swipe readers for call access and charging. Such technology can be utilised as a convenient payment system for on-board retailing. The satellite communications used in providing an on-board telephone facility may also offer passengers the possibility of accessing the Internet and even placing orders for merchandise from the TV home shopping channels such as QVC and HSN.

Such developments pose potentially serious challenges to today's model of airport retailing and will require a proactive response on the part of both the airport operators and the retailers. Some airport authorities, such as Aer Rianta, have made movements in this direction with the introduction of the Dublin Airport Executive Club. Membership of this club provides special car parking close to the terminal, a departure lounge which has full business centre facilities, automatic membership of the airport wine club, and special discounts and exclusive offers in the airport shops. The service has the potential to evolve into the provision of special shops exclusive to members, with unique product ranges and special branding.

The customer loyalty schemes introduced by Scandinavian Airline Services and the BAA involve the use of the latest information technologies.

However, evidence of widescale attempts to market to the travelling consumer prior to and during their journey is still not in evidence. Technology provides both the airlines and airports with the opportunity to target potential and actual passengers from the time the decision to travel has been made, throughout their journey and subsequently at their place of arrival. Better database management techniques combined with the growing sophistication of direct mail systems and telecommunications provide the opportunity to communicate with customers and overcome the pressures associated with time constraints within the airport.

The structural changes occurring within the market, combined with increased competitive pressures, may compel some duty/tax-free retailers to be more proactive in developing their merchandise mix. As illustrated in Chapter 4, the main impetus for new product development in duty/tax-free retailing has come from suppliers rather than retailers, and has been focused largely in the alcohol and perfume areas. If the price advantage associated with airport shopping is removed, then there will be a compulsion upon retailers also to readjust their marketing mix and to explore the possibilities of achieving differentiation through the uniqueness of the product offer rather than the price.

Some of the development work undertaken by suppliers, such as World Brands, has centred on generating demand for products designed exclusively for the duty/tax-free market. Such a strategy creates a degree of differentiation and it remains logical to expect that such offers would remain part of airport retailing even if duty/tax-free was abolished. Special packaging, unique formulations and vintages, coupled with exclusive designs, may offer the airport retailer some possibility of maintaining a market distinct from that of the domestic trader. Such a strategy will require airport operators and retailers to adopt a co-operative rather than confrontational approach with suppliers. By replicating the practices in other retail sectors and by working together as a partnership, all parties are provided with the opportunity to jointly develop new products or line extensions.

Airports have the option of making their highly detailed passenger and customer data lists available to such a partnership. This could be coupled with the market research data gathered on a regular basis at most airports. If combined with manufacturers' and suppliers' comprehensive knowledge of consumer tastes and behaviours, such an information system would allow the development of products tailored to the tastes of specific airport customer segments.

Airport retailers have the opportunity to derive considerable trading benefits through such collaboration. While some test marketing is conducted within the airport environment, the possibility exists for a far greater number of new product concepts to be trialed prior to their launch on to national markets. The exclusive availability of the latest products at airports would stimulate demand among passengers and provide the airport retailer with premium pricing opportunities with no domestic comparisons.

As the main airports expand and passenger volumes increase, the potential shopping population will become greater than in many large cities. This will allow further segmentation of airport retailing and provide viable market opportunities for a range of niche retail offers, including antiques, art, specialist books, philately, gourmet foods and drinks. More companies are coming to realise the potential of airports and are seeking to join major retailers such as Virgin, Dixons, Liberty, The Gap and leading fashion names such as Hermes, Gucci and Ferragamo. Further segmentation of the retail offer therefore remains a likely strategy for the airport authorities.

The growth of speciality-based retailing used so effectively in European airports such as Copenhagen, Heathrow, Madrid, Vienna and Milan and in Middle Eastern airports such as Abu Dhabi, is likely to become more widespread. Such a strategy will allow many airport operators to develop a contingency plan and reduce their reliance on a limited product base. The BAA, for example, has introduced a programme of expansion at its terminals designed specifically to provide additional quality retail space. Abu Dhabi is using its new and enlarged range of high-quality specialist shops to establish a distinctive identity that clearly separates it from its high-profile Gulf neighbours.

In order to offer a quality shopping experience, airport operators in the future will be required to integrate the retail environment within the overall design of the airport terminal. The marginalisation of retail operations in the design and construction of terminal facilities will no longer be an acceptable option for an airport wishing to maximise its income-generating capabilities. The central role that retailing plays in the generation of airport revenues has firmly established it as a key service that has to be provided. Despite the protestations of some within the aviation industry, the importance of retail income to the successful operation of an airport has led to commercial activities being represented at the very highest levels of airport management.

Secondary hubs will continue to mirror the trends of the larger airports and seek to attract a range of retail brands designed to augment their existing duty/tax-free product offer. The growth in passenger numbers stimulated by the liberalisation process will enable many of these secondary airports to achieve the critical mass needed to support a credible and economically viable, speciality retail mix. Retail offers based on merchandise from the locality or featuring country-typical goods will continue to play an important role in establishing the uniqueness of these airports.

The smaller regional airports will continue to be faced with the challenge of trying to expand their revenue base in the face of continuing downward pressure on aeronautical charges. The core duty/tax-free offer will remain the dominant feature of retailing at these airports with economies of scale limiting efforts to introduce niche and speciality merchandise. In the light of this dependency, the proposal to abolish intra-EU duty/tax-free sales will have the greatest impact at the regional level. Provincial airports will be faced with a significant challenge to their ability to create a credible retail offer and to maximise their revenue-earning potential from commercial activities. One possible option may be to offer the airport as a total retail package to an interested party, who in return might guarantee the airport a fixed fee per passenger. Alpha Retail Trading, for example, has the potential to provide a portfolio of specialist services, including catering, cosmetic centres, drug stores, book stores as well as duty/tax-free retailing.

As air travel continues to expand and passengers become familiar with airports, operators will be faced with finding new ways of ensuring that the retail offer remains appealing to consumers. The US concept of retail being theatre may offer possibilities for the airport authorities. Disney or Warner Bros. stores have the potential to add an entertainment dimension to the retailing mix at the airport. Leisure activities for passengers in transit, such as sport and fitness facilities, cinemas and family entertainment centres, also offer the opportunity for a lifestyle-based retail provision.

The creation of retail and leisure facilities on the landside of an airport may provide a contingency strategy for operators faced with the possible demise of intra-EU duty/tax-free sales. At airports in the Far East, for example, golf courses and themed restaurants have been successfully developed and there remain few reasons why a similar provision could not be offered within Europe. Gatwick airport, for example, has developed a series of activities centred upon aviation; for instance, a flight simulator within the terminal building has become a significant revenue generator.

While the shopping mall adjacent to Schiphol airport remains an anomaly in the context of European airport retailing, a strategy of targeting the non-travelling consumer opens a wide range of both retailing and service-based opportunities. Gatwick's development of a theme park in the landside environs of the airport illustrates the potential opportunities that can stem from pursuing an alternative consumer base. In the same way that certain regional airports have been described as industrial estates with runways attached, then one may expect many hub airports to become shopping complexes and leisure centres located at inter-modal junctures.

While many retailers have focused upon achieving operational efficiencies within their stores or at head office, the logistics function in many companies has remained a neglected entity. The importance of logistics and supply chain management for airport retailing was illustrated in Chapter 5. The transient nature of the airport customer base, with passengers having only one chance to shop, means that merchandise 'stock-outs' represent permanently lost sales.

Equally, the high volumes of product turnover associated with duty-free retailing means that fast and cost-effective store restocking is a critical service function. Any out-of-stocks resulting from a failure to deliver will have revenue implications for the retailer and a loss of concession income for the operator. Investment in technology has improved the flow of product by providing better information and prompting organisational efficiencies. However, in comparison with other industrial sectors and in relation to other areas of retailing, airport retail logistics are not as advanced. For example, there is only limited evidence of the use of EDI or JIT operations in duty free. The most sophisticated systems tend to be operated by domestic retailers whose duty-free business is an adjunct to their high street operations.

While Customs and Excise bond regulations will undoubtedly influence the logistics function within an airport, opportunities exist to improve operational efficiencies within the supply chain. Some airport authorities, such as the BAA, have experimented with vertical integration and begun to develop their own brands of liquor. Others view the acquisition of Duty Free Shoppers (DFS) by LVMH, the luxury goods manufacturing group, as a move towards the concept of a total vertical marketing system.

Another way of improving operational efficiencies is through greater supply chain co-operation. The existence of an active airport operator/retailer/manufacturer partnership could exploit current data transmission technologies to enable real-time sales volumes and product data to be transferred from point of sale to manufacturing plants or supplier warehouses. As noted above, the threat of duty/tax-free abolition has provided an incentive for closer collaboration in new product development. Such co-operation would also facilitate production and logistics planning, improve the merchandise selection process and make JIT a real possibility. There remains little evidence however of the existence of contractual channels or vertical co-ordination between the major operators within the market. While many relationships between retailer and supplier are long standing, there is little to suggest any form of alliance. Contacts with suppliers are limited to contract negotiations and monitoring and supply issues. The operation of the supply chain has therefore more to do with the issue of countervailing power than any adherence to mutuality and partnership.

Many organisations outside the air transport industry have realised the potential impact upon service and efficiency that a well co-ordinated supply chain can have. Given the highly competitive nature of airport retailing, then one may argue that improvements within the supply chain are not only desirable but are of fundamental importance to the future survival of airport retailers. Those who embark upon a strategy of partnership, co-operation and information sharing are likely to be those who establish themselves as having a distribution differentiation over their competitors. In the longer term, as price loses its distinguishing value, it is the collaborative process with suppliers that will enable airport retailers to position themselves as being different from the high street trader. Such an approach will provide airport shops with internationally branded products that are exclusive, unique and valued by the consumer.

Central to the concept of value delivery within the exchange process are the notions of supportive HRM strategies and customer care techniques. Such skills have become increasingly important for a number of reasons. First, consumers have become more discerning and critical. With experience gained from modern shopping malls, outlet parks and high street chains, similar levels of service and choice are expected from airport stores. Secondly, travel is no longer confined to a select sub-set of the population, with passengers now being drawn from all social classes. Finally, non-travelling consumer groups are beginning to use airports with much greater frequency. Retailers therefore have to cope with increasing social diversity.

Many airport retailers have sought to respond to consumer change through improvements to their operational procedures, for example by accepting a wider range of currencies and credit cards, and stocking culturally specific merchandise. While undoubtedly important, such a strategy in itself will be insufficient. The skill requirement of those individuals working in store will also have to increase. In international airports especially, employees will be expected to be multi-lingual, have an appreciation of cultural traits and tendencies, and an ability to work long, unsociable hours. While examples do exist, staff training to this level of proficiency is not widespread and as such represents an exception to the rule rather than an industry norm. Allders, the duty/tax-free retailer, for example, has continued to benefit from the employment of trained Japanese personnel in its store in Heathrow's Terminal 4. The presence of bi-lingual Japanese nationals, staffing a dedicated sales area, has a powerfully beneficial effect on sales and provides the retailer with a discernible differentiating factor.

Given the demands of an airport environment, one may expect the emphasis to be placed on selecting highly trained individuals with good social skills and expert product knowledge. Significantly, however, there remain wide variations in the way in which airport retailers conduct their HRM strategies. Differences exist in the prerequisites asked for by employers, the process of recruitment undertaken and the training available to employees. In many ways this is unsurprising and a reflection of the retail sector as a whole. While these variances may be characteristic of retailing generally, given the competitive pressures operating within the airport trading environment, retailers do not have the luxury of losing sales through inadequately skilled staff.

As already noted, passengers represent a one-off sales opportunity and display unique behavioural characteristics during the travel process. Training is therefore required which engenders an understanding of passenger psychology, develops excellent product knowledge, reduces customer dissonance and instils a sense of security in the transaction. Because of the nature of the airport trading environment, a large proportion of employees work full time. This provides retailers with the opportunity to develop personnel through their own internal labour market. The criticism that training represents a cost rather than an investment would be difficult to sustain, as employees are sheltered from the economic demands of the general labour market and tied in closer to the individual organisation.

In conclusion, therefore, an understanding of airport retailing cannot be divorced from the wider regulatory and competitive imperatives that are a function of the air transport industry. The evolution of the market is set to continue, with South America, Asia and parts of Africa emerging as potential revenue streams for European companies looking to expand operations and overcome the potential limitations of competition and duty/tax-free abolition. The complexity of the market is epitomised by the alliances, mergers and takeovers that are occurring on a global scale. One may therefore characterise the air transport industry as a dynamic and growing sector that will continue to serve as a catalyst for continued retail expansion.

Glossary

ADP Aeroports de Paris, the authority responsible for the development and operation of the airports of Paris – CDG One and Two, Orly Sud and Orly Ouest.

AENA The Spanish airport authority, responsible for the development and operation of all state-owned airports in Spain and its associated territories.

Aeronautical revenues All revenues earned by the airport from aviation activity including landing charges, passenger load fees, airbridge fees, cargo throughput, rents for check-in desks and offices, aviation fuel throughput charges and the provision of security.

Airbridge/Jetway A movable telescopic tunnel linking the departure or arrival gate to the aircraft doorway. The airbridge may be hired or lowered to suit different types of aircraft.

Airbridge fees A fixed fee which an airline pays to the airport for the use of an airbridge or jetway. The charge may be related to the length of time the aircraft occupies the stand.

Airfield That part of the aerodrome or airport to be used for take-off, landing and taxiing of aircraft, consisting of the manoeuvring area and the apron(s), i.e. the large areas of concrete between the terminal buildings and the taxiways and runways.

Airside (terminal) The departure and arrival areas of the terminal building beyond security, passport and Customs controls. Access is restricted to travelling passengers with valid boarding cards, airline, airport and other authorised staff. Departure level airside normally includes those passenger lounges, shopping areas and departure gates which are accessible only through security or immigration control. Arrival level airside includes arrival gates, passport controls and baggage collection areas prior to exiting into the public area.

Apron Airside area for parking and manoeuvre of aircraft and ground-handling equipment. It does not include taxiways.

Apron services (*see* **Ground handling**) taxiways, runways and in the air.

BAA BAA plc, the British Airport Authority.

Baggage handling The processing of passenger baggage from check-in to aircraft to arrivals.

CAA The Civil Aviation Authority of the UK responsible for the regulation of the aviation industry there and the issuing of operating licences to airlines and aircraft owners.

Charter passenger flights Non-scheduled flights on which the seats are block booked by a tour operator or travel agent and normally sold as part of a complete holiday package.

Chicago Convention A conference on the regulation of post-war air transport held in Chicago in 1944, where *inter alia*, the basis for today's duty/tax-free passenger allowances was laid.

Commercial revenues   All revenues earned by the airport authority from non-aviation sources, including concession fees from shopping, banking, catering and bars, car parking and advertising sign rentals.

Deregulation (US)   The internal US air transport market became fully liberalised in 1978 with all routes opened up for competitive access. This liberalisation, or deregulation as it was called in the USA, became the model for liberalising the aviation market in Europe.

Duty free   Goods purchased in a duty-free shop which are normally subject to excise duties and VAT.

Eurostar   Brand name for rapid train (though not yet a TGV) used on the Channel Tunnel route between the UK and continental Europe.

Ground-handling/Apron services   The processing of an airline's passengers through check-in, baggage delivery and embarkation, both at departure and on arrival. Responsibilities may include load control, surface transport and security documentation, the servicing of the aircraft on the ground (apron services) including cleaning, catering, refuelling, marshalling on to the stand or parking place, and dealing with the airport authority.

Hubbing/Hub and spoke   Airport hub and spoke systems were developed by US airline carriers as an efficient means of serving the large US domestic air traffic market following deregulation. The underlying philosophy is that there are routes which alone are not viable, e.g. Providence, Rhode Island to Austin, Texas, but when combined with other routes through a single hub airport, e.g. Providence/Cincinnati to Dallas/Austin, can produce economic passenger loads. The formula which defines the marketing power of a hub is $N(N-1)/2$ where N equals the number of spokes. At Dallas, for example, there may be 50 spokes which, using the formula, produces over 1200 possible routings. In Europe the principal hub airports are London Heathrow and Gatwick, Amsterdam Schiphol, Frankfurt, Paris CDG, Copenhagen, Rome and Zurich.

IATA   International Air Transport Association, the airlines international trade association.

ICAO   International Civil Aviation Organisation, the international regulatory body for civil aviation.

Inter-lining   Air journeys involving a number of stages or sectors and using different airlines for some of the stages.

Inter-modal   The node or connection point at which it is possible to move from one mode of transport to another, for example, at an airport which has a train station and a bus terminus.

Intra-EU/Intra-Community   Journeys between two or more member states of the European Union.

Landing charges   These are the charges levied by the airport on an aircraft for the use of the runways and taxiways, and are normally applied on an aircraft landing. In general, landing charges are based on the weight of the aircraft.

Landside   All non-operational and public areas of the airport and terminal buildings used for passenger access, car parking, check-in, shopping, restaurants and bars, and general waiting areas.

Liberalisation   The liberalisation of the air transport industry in the European Union, completed in April 1997, allows all EU airline carriers holding an operating licence free access to all international routes within the EU, subject to slot availability. The airlines are largely free to charge whatever fares they wish.

Load fee   (*see* **Passenger charges**)

Long-haul flights   A descriptive term generally applied to intercontinental flights.

Lux levels   A unit of illumination. Used to indicate the level of light in a defined space.

Master planning   An airport master plan is the overall development plan for an airport and the corresponding land use surrounding the airport. It includes the physical outline of the proposed development and describes the phasing, the financial implications and implementation strategies involved.

MTOW   Maximum take-off weight of an aircraft as certified by the manufacturer.

Navigational infrastructure   The system of visual and radio navigational aids used in the guidance of aircraft on the ground and in the air.

Passenger charges/Load fee   The passenger load fee or charge is paid to the airport by the airline, for each passenger, and is included in the ticket price. In most places the charge is not shown separately. The charge relates to the airport services provided for passengers, e.g. use of the terminal building, customs and passport controls, and security.

Passenger manifesto   A detailed breakdown of the composition of the passengers on-board an aircraft. This may include the number of first and business class passengers, the number of economy class travellers and the number of unaccompanied minors.

Pier   A fixed corridor which links the terminal to an aircraft stand or busing gate.

Runway capacity   Number of aircraft movements, in and out, which can be operated safely on the runway, when measured against an acceptable delay criterion.

Runway congestion   When demand on the runway system exceeds capacity, and approaching or departing aircraft have to hold and suffer delays.

Scheduled passenger flights   Flights scheduled, performed according to a published timetable, or so regular or frequent as to constitute a recognisably systematic series, which are open to use by the public on an individually ticketed basis.

Schengen Agreement   A treaty, the object of which is to provide for the free movement of people between and within signatory states. All EU member states, except the UK and Ireland, are signatories along with Norway and Iceland. Because there is free movement between states, non-Schengen passengers must be kept in a separate area in the airports involved. This requirement has implications for airport capacity and demands the duplication of passenger and shopping facilities.

Slots   The designated time of arrival and departure of an aircraft at an airport.

Tax free   Goods purchased in a tax-free shop which are normally subject to Value Added Tax only.

Taxiing   The ground handling of aircraft in transit to or from apron areas, or to or from runways.

Terminal   Building(s) used for the processing, embarking and disembarking of passengers.

Terminal capacity   The number of passengers who can be processed in the terminal building measured against assumed standards of safety and comfort.

TGV   Train Grande Vitesse – the French rapid train system which can operate at speeds in excess of 240 km per hour.

Transit passengers   Passengers on aircraft who make a stop at an airport for regulatory, technical or operational reasons, and continue their journey on the same aircraft.

Vendor control   A system agreed by EU member states to regulate the sale of tax/duty-free goods in the absence of customs barriers within the Single Market. It places the responsibility upon duty/tax-free retailers to ensure that air and sea passengers do not purchase goods in excess of their given allowances.

Bibliography and References

ACI (1992)   *The Economic Benefits of Air Transport*, Air Transport Action Group, IATA, Switzerland.

ACI Europe (1996)   *Economics Committee Survey*, Airports Council International, Brussels.

Air Transport Group (1997)   *The Impact of the Loss of Intra-EU Duty and Tax Free Sales on Airport Development Economics*, College of Aeronautics, Cranfield University, UK.

Alexander, N. (1997)   *International Retailing*, Blackwell, Oxford.

Ambler, T. (1992)   'The role of duty free in global marketing', *Business Strategy Review*, Autumn, pp. 57–72.

Anderson, P. (1982)   'Marketing, strategic planning and the theory of the firm', *Journal of Marketing*, 46 Spring, pp. 16–26.

Ansoff, I. (1987)   *Corporate Strategy*, revised edn, Penguin, London.

Arndt, J. (1983)   'The political economy paradigm: foundation for theory building in marketing', *Journal of Marketing*, 47, Fall, pp. 44–54.

BAA (undated)   *The Case for Terminal 5: Facing up to Britain's 21st Century Airport Needs*, BAA, London.

BAA (1995)   'Duty free bargains, take off', *BAA Staff Magazine*, October, p. 1.

BAA (1996)   *Annual Report and Accounts*, BAA, London.

Baden-Fuller, C. (1986)   'Rising concentration: the UK grocery trade 1970–1980', in Tucker, K. and Baden-Fuller, C. (eds), *Firms and Markets*, Croom Helm, London, pp. 63–82.

Ballini, P. (1993)   'The Customers' Expectations – Quality Measurement', paper presented at the 2nd Airports Council International Conference, Milan.

Barr, V. and Broudy, C. (1986)   *Designing to Sell: A Complete Guide to Retail Store Planning and Design*, McGraw-Hill, New York.

Becker, G. (1964)   *Human Capital*, National Bureau of Economic Research, New York.

Beier, F. and Stern, F. (1969)   'Power in the channels of distribution', in Stern, W. (ed.), *Distribution Channels: Behavioural Dimensions*, Houghton Mifflin, Boston, MA.

Best 'n' Most (1996)   Generation AB Publications, Ornskoldsvik, Sweden.

Bingman, C. (1996)   'Airports as Retail Malls? What are the Criteria for a Successful Airport Retail Mall?, paper presented at 'Airports – A Major Trading Opportunity', Airports Council International Conference, Marseille.

Bipe Conseil (1997)   *Relative Importance of Duty and Tax Free and Explicit State Subsidies on Competition between Rail and Air Travel on Intra-EU Routes*, Bipe Conseil, Paris.

BMS (1994)   *The Brand Equity of UK Regional Airports*, Business Marketing Services (mimeo).

Bowlby, S. and Foord, J. (1995)   'Relational contracting between UK retailers and manufacturers', *International Review of Retailing, Distribution and Consumer Research*, 5(3), pp. 333–360.

Brendal, G. (1994)   'Airport retail pricing – the critical factor or: how to keep customers and win new ones', *Commercial Airport 1994/95*, Stirling Publications, London.

Broadbridge, A. (1995)   'Female and male earnings differentials in retailing', *Service Industries Journal*, 15(1), pp. 14–34.

Broadbridge, A. (1996)   'Female and male managers – equal progression?', *International Review of Retailing, Distribution and Consumer Research*, 6(1), pp. 259–279.

Broadbridge, A. (1998) 'Barriers in the career progression of retail managers', *International Review of Retailing, Distribution and Consumer Research*, 8(1), pp. 53–78.

Bromley, R. and Thomas, C. (1993)   'The retail revolution, the carless shopper and disadvantage', *Transactions of the Institute of British Geographers*, 18, pp. 222–236.

Brown, S. (1995)   *Postmodern Marketing*, Routledge, London.

Brownlie, D. (1985)   'The anatomy of strategic market planning', *Journal of Marketing Management*, 1, pp. 35–63.

Campbell, J. (1994)   'Commentary: tobacco products. Changes fire the tobacco giants', *Duty Free Data Base and Directory*, Raven Fox, London, pp. 129–131.

Cathelat, B. (1990)   *Socio-Styles*, Kogan Page, London.

Chesterton (1994)   *Airport Retailing: The Growth of a New High Street*, Chesterton International Property Consultants, London.

Cooper, J., Browne, M. and Peters, M. (1994)   *European Logistics: Markets, Management and Strategy*, Blackwell, Oxford.

Corporate Intelligence Group (1997)   *Airport Retailing in the UK*, Corporate Intelligence Group, London.

Cranfield (1997)   *Airport Development Economics: A Study for the European Travel Research Foundation*, Air Transport Group, Cranfield College of Aeronautics, UK.

Crosier, K. (1994)   'Promotion', in Baker, M. (ed.), *The Marketing Book*, 3rd edn, Butterworth-Heinemann, Oxford.

Dawson, J. (1995)   'Retail change in the European Community', in Davies, R. (ed.), *Retail Planning Policies in Western Europe*, Routledge, London.

de Man, W. (1996)   'Pleasing the Airport Customer', paper presented at 'Good Communication for Better Airport Marketing', ACI Europe Conference, Bologna.

Denning, J. and Freathy, P. (1996)   'Retail strategies for the petrol forecourt: the example of RoadChef Forecourts Ltd', *International Review of Retailing, Distribution and Consumer Research*, 6(1), pp. 97–112.

Dex, S. (1988)   'Gender and the labour market', in Gallie, D. (ed.), *Employment in Britain*, Blackwell, Oxford.

DFNI (1997)   'Survey of airport and ferry travellers' purchasing habits', *Duty Free News International*, London, March 15–31, pp. 60–63.

Dickinson, R. and Hollander, S. (1996)   'Some definition problems in marketing channels', *Journal of Marketing Channels* 5(1), pp. 1–15.

Doeringer, P. and Piore, M. (1971)   *Internal Labour Markets and Manpower Analysis*, D. C. Heath, Massachusetts.

Doganis, R. (1992)   *The Airport Business*, Routledge, London.

Doganis, R. (1993) 'Economic Issues in Airport Management', paper presented at the Airport Economics and Finance Symposium, Department of Air Transport, Cranfield University, UK.

Doganis, R. (1995) 'Airport Economics – Some Fundamental Principles', paper presented at Economics and Finance Symposium, University of Westminster and Cranfield University, UK.

Doganis, R., Lobbenberg, A. and Graham, A. (1994) *A Comparative Study of Value for Money at Australian and European Airports*, Summary Report for Participating Airports (mimeo).

Drucker, P. (1954) *The Practice of Management*, Butterworth-Heinemann, Oxford (reprint 1993).

Duty Free Data Base and Directory (1994)	Raven Fox, London.

Duty Free Data Base and Directory (1996/97)	Raven Fox, London.

Dwyer, F., Schurr, P. and Oh. S. (1987) 'Developing buyer–seller relationships', *Journal of Marketing*, 51, pp. 11–27.

EIRR (1997) 'Working in Europe', *European Industrial Relations Review*, 282, pp. 16–21.

El-Ansary, A. and Stern, L. (1972) 'Power measurement in the distribution channel', *Journal of Marketing Research*, 9, pp. 47–52.

Emerson, R. (1962) 'Power-dependence relations', *American Sociological Review*, 27, pp. 31–41.

ETRF (1996) *Facts about the Duty and Tax Free Industry in the EU: 1995 Statistics*, European Travel Research Foundation, Surrey.

Eurostat (1994) *Retailing in the European Economic Area*, European Commission Statistical Office, Luxembourg.

Fernie, J. (1992) 'Distribution strategies of European retailers', *European Journal of Marketing*, 26(8/9), pp. 35–47.

Fernie, J. (1995) 'International comparisons of supply chain management in grocery retailing', *Service Industries Journal*, 5, pp. 134–147.

Freathy, P. (1993) 'Developments in the superstore labour market', *Service Industries Journal*, 13, pp. 65–79.

Freathy, P. (1997) 'A Typology of Airport Retailing', paper presented at the 9th International Conference on Research in the Distributive Trades, Leuven, Belgium, pp. C5.20–C5.29.

Freathy, P. and Sparks, L. (1994) 'Contemporary developments in employee relations in food retailing', *Service Industries Journal*, 14, pp. 499–514.

Freathy, P. and Sparks, L. (1995) 'Flexibility, labour segmentation and retail superstore managers: the effects of Sunday trading', *International Review of Retail, Distribution and Consumer Research*, 5, pp. 361–385.

French, J. and Raven, B. (1959) 'The basis of social power', in Cartwright, D. (ed.), *Studies in Social Power*, University of Michigan Press, Ann Arbor, MI.

Furlong, A. (1990) 'Labour market segmentation and the age structuring of employment opportunities for young people', *Work Employment and Society*, 4(2), pp. 253–270.

Gaski, J. (1984) 'The theory of power and conflict in channels of distribution', *Journal of Marketing*, 48(3), pp. 9–29.

Gaski, J. and Nevin, J. (1985) 'The differential effects of exercised and unexercised power sources in a marketing channel', *Journal of Marketing Research*, 22, pp. 130–142.

Gibson, B. (1992) 'The Role of High Street Names in Airport Trading', paper presented at Airports Association Council International, First Airport Trading Conference, Paris.

Gray, F. (1994) 'Trends in Airport Retail Property – Consequences for Space Allocation, New Development and Values', paper presented at the Airport Associated Property Conference, Henry Stewart Studies, London.

Gray, F. (1997) 'Getting the Process Right in Airports Large and Small – External Help – Concession Planning', paper presented at the 6th World Airport Trading Conference and Exhibition, Lisbon.

Greenley, G. (1982) 'An overview of marketing planning in UK manufacturing companies', *European Journal of Marketing*, 16(7), pp. 3–15.

Greenley, G. (1986) *The Strategic and Operational Planning of Marketing*, McGraw-Hill, London.

Harris, D. and Walters, D. (1992) *Retail Operations Management*, Prentice-Hall, Hemel-Hempstead, UK.

Harrison, T. (1995) 'Segmenting the market for retail financial services', *International Review of Retail, Distribution and Consumer Research*, 5, pp. 271–286.

Hooley, G. and Saunders, J. (1993) *Competitive Positioning: The Key to Market Success*, Prentice-Hall, Hemel-Hempstead, UK.

Humphries, G. (1996) *The Future of Airport Retailing: Opportunities and Threats in a Global Market*, Financial Times Management Report, London.

Hunt, S. and Nevin, J. (1974) 'Power in a channel of distribution: sources and consequences', *Journal of Marketing Research*, 11, pp. 186–193.

Hunter, L. and Mulvey, C. (1981) *Economics of Wages and Labour*, Macmillan, London.

ICAA (1990) *'Economic Impact of the Potential Loss of Intra-EC Duty and Tax Free Sales on the Air Transport Industry'*, report prepared by Coopers and Lybrand, London.

IDFC (1989) 'Duty Free within the Single Internal Market: A Position Paper', unpublished monograph, International Duty Free Confederation.

IRS (1997) *An Assessment of the Impact of the Abolition of Intra-EU Duty and Tax Free Allowances*, Institute for Retail Studies, University of Stirling, UK.

Keith, R. (1960) 'The marketing revolution', *Journal of Marketing*, 24, January, pp. 35–38.

Keogh, D. (1994) 'An airport authority with a difference – the Irish experience', *Commercial Airport 1994/95*, Stirling Publications, London.

Kerr, C. (1954) 'The Balkanisation of labour markets', in Bakke, E. (ed.), *Labour Mobility and Economic Opportunity*, John Wiley, New York.

Klapper, S. (1995) 'Innovative Use of Mobile Sales Units', paper presented at 4th World Airport Trading Conference, London.

Kollatt, D., Blackwell, R. and Robeson, J. (1972) *Strategic Marketing*, Holt Reinhart & Winston Inc, New York.

Kotler, P. (1991) *Marketing Management Analysis, Planning, Implementation and Control*, 7th edn, Prentice-Hall, Englewood Cliffs, NJ.

Labour Studies Group (1985) 'Economic, social and political factors in the operation of the labour market', in Roberts, B., Finnegan, R. and Gallie, D. (eds.), *New Approaches to Economic Life: Economic Restructuring and the Social Division of Labour*, Manchester University Press, Manchester.

Leighton, D. (1966)   *International Marketing: Text and Cases*, McGraw-Hill, New York.
Lester, R. (1952)   'A range theory of wage differentials', *Industrial and Labour Relations Review*, 5, July.
Levitt, T. (1960)   'Marketing myopia', *Harvard Business Review*, 38, July–August, pp. 45–56.
Levitt, T. (1986)   *The Marketing Imagination*, Free Press, New York.
Livingstone, B. (1995)   'Partnership in duty free', in Proceedings of the 4th World Trading Conference and Exhibition, London.
Lloyd-Jones, T. (1996)   'Duty free market sends out mixed signals', *Duty Free Data Base and Directory 1996/97*, Raven Fox, London.
Lundqvist, G. (1994)   'Is there life after duty free?', *Commercial Airport 1994/95*, Stirling Publications, London.
Lusch, R. (1982)   *Management of Retail Enterprises*, Kent Publishing, Boston, MA.
McDonald, M. (1994)   'Developing the marketing plan', in Baker, M. (ed.), *The Marketing Book*, 3rd edn, Butterworth-Heinemann, Oxford.
McDonald, M. (1995)   *Marketing Plans: How to Prepare Them; How to Use Them*, 3rd edn, Butterworth-Heinemann, Oxford.
McDonald, M. (1996)   'Strategic marketing planning: theory, practice and research agendas', *Journal of Marketing Management*, 12, pp. 5–27.
McGoldrick, P. (1990)   *Retail Marketing*, McGraw-Hill, London.
Michon, F. (1987)   'Segmentation, employment structures and production structures', in Tarling, R. (ed.), *Flexibility in Labour Markets*, Academic Press, London.
Mintel (1997)   *Duty/Tax Free Retailing*, Mintel Retail Intelligence, London.
MMC (1996)   *A Report on the Regulation of the London Airport Companies (Heathrow Airport Ltd, Gatwick Airport Ltd, Stanstead Airport Ltd)*, Monopolies and Mergers Commission, London.
Morrison, P. (1990)   'Segmentation theory applied to local regional and spatial labour markets', *Progress in Human Geography*, 14(4), pp. 488–528.
NatWest Securities Corporation (1996)   *Company Report on BAA*, NatWest Securities, London.
Netherlands Economic Institute (1989)   *The Impact of Abolishing Duty and Tax Free Allowances in the European Community*, Department for Society and Policy, Rotterdam.
O'Connell, F. (1993)   'An Aer Rianta Response to Aspects of Change in the Micro, Task and Organisation Environments', unpublished mimeo, Aer Rianta, Dublin.
Ogbonna, E. and Wilkinson, B. (1996)   'Inter-organizational power relations in the UK grocery industry: contradictions and developments', *International Review of Retail Distribution and Consumer Research*, 6, pp. 395–414.
O'Toole, K. (1997)   'European lead: Europe's hubs begin to face up to the new commercialism', *Flight International*, 151(4572), p. 34.
PA Consulting (1996)   Unpublished report produced by Howarth & Howarth in 1989 for Maltese Government and updated by PA Consulting, London.
Papagiorcopulo, G. (1994)   'The importance of non-aviation revenues to a small airport', *Commercial Airport 1994/95*, Stirling Publications, London.

Peck, J. (1988) 'The Structure and Segmentation of Local Labour Markets: Aspects of the Geographical Anatomy of Youth Unemployment in Great Britain', unpublished PhD thesis, University of Manchester.

Peck, J. (1989) 'Reconceptualising the local labour market: space segmentation and the state', *Progress in Human Geography*, 13(1), pp. 42–61.

Peck, J. (1996) *Work-Place: The Social Regulation of Labor Markets*, Guilford Press, New York.

Piore, M. (1975) 'Notes for a theory of labour market stratification', in Edwards, R., Reich, M. and Gordon, D. (eds), *Labour Market Segmentation*, Lexington Heath, Massachusetts.

Porter, M. (1980) *Competitive Strategy*, Free Press, New York.

Portland Group (1996) *A Review of Airport Privatisation*, report for Aer Rianta, Portland Group, London.

Pyke, F. (1986) *Labour Flexibility and the Use of Time*, CURID, Department of Geography, University of Manchester.

Pyke, F. (1988) *Local Labour Markets and the Organisation of Time: Reflections on the Rise of Part-Time Working*, CURID, Geography Department, University of Manchester.

Robinson, T. and Clarke-Hill, C. (1995) 'International alliances in European retailing', *International Review of Retail, Distribution and Consumer Research*, 5(2), pp. 167–184.

Rubery, J. (1978) 'Structured labour markets, worker organisation and low pay', *Cambridge Journal of Economics*, 2(1), pp. 17–36.

Rushton, A. (1982) '*The Balance of Power in a Marketing Channel*' paper presented at ESOMAR Seminar, Profitable Co-operation of Manufacturers and Retailers – The Contribution of Research, ESOMAR, Brussels.

Sachdev, H., Bello, D. and Verhage, B. (1995) 'Export involvement and channel conflict in a manufacturer–intermediary relationship', *Journal of Marketing Channels*, 4(4), pp. 37–63.

Schiller, R. (1986) 'Retail decentralisation and the coming of the third wave', *The Planner*, 72(7), pp. 13–15.

SD10 (1975) *Business Monitor: Report on the Census of Distribution and Other Services 1971*, HMSO, London.

SDA25 (1996) *Service Sector: Results of the 1994 Retailing Survey*, HMSO, London.

Sewell-Rutter, C. (1995) 'Experiences of Privatisation', paper presented at the Airport Economics and Finance Symposium, University of Westminster/Cranfield University, UK.

Shaw, R. (1993) 'Airport Retailing', paper presented at Airport Associated Property Conference, Henry Stewart Studies, London.

SH&E (1997) *The Impact of the Abolition of intra-EU Duty and Tax Free Sales on the Economics of Charter Airlines*, report commissioned by the European Travel Research Foundation, London SH&E.

Shipley, D. (1987) 'What British distributors dislike about manufacturers', *European Journal of Marketing*, 21(3), pp. 77–88.

Siguaw, J. and Hoffman, D. (1995) 'The effects of distribution fees on channel relationships within the USA', *International Review of Retail Distribution and Consumer Research*, 5, pp. 21–35.

Simon, N. (1993) 'Liquor and Tobacco – The Future', paper presented at the 2nd Airports Council International Conference, Milan.

Slater, A. (1990) 'The impact of information technology', in Fernie, J. (ed.), *Retail Distribution Management*, Kogan Page, London.

Slichter, S. (1950) 'Note on the structure of wages', *Review of Economics and Statistics*, February.

Smith, C. (1994) *Airport Industry Structure: The Trend Towards Commercialization*, report prepared by Coopers and Lybrand, London.

Smith, D. and Sparks, L. (1993) 'The transformation of physical distribution in retailing: the example of Tesco plc', *International Review of Retailing, Distribution and Consumer Research*, 3 pp. 35–64.

Sparks, L. (1991) 'Retailing in the 1990s: differentiation through customer service?', *Irish Marketing Review*, 5, pp. 28–38.

Sparks, L. (1993) 'The rise and fall of mass marketing? Food retailing in Great Britain since 1960', in Tedlow, R. and Jones, G. (eds), *The Rise and Fall of Mass Marketing*, Routledge, London.

Stannack, P. (1996) 'Purchasing power and supply chain management power – two different paradigms? – a response to Ramsay's 'Purchasing power'', *European Journal of Purchasing and Supply Management*, 2, pp. 47–56.

Stephenson, D. and Willett, R. (1969) 'Analysis of customers' retail patronage strategies', in McDonald, P. (ed.), *Marketing Involvement in Society and Economy*, AMA, Chicago, pp. 316–322.

Stern, L. and El-Ansary, A. (1988) *Marketing Channels*, Prentice-Hall, Englewood Cliffs, NJ.

Symons, Travers, Morgan (1997) *Assessment of the Impact of the Abolition of Intra-EU Duty and Tax Free Allowances on Low Cost Airlines*, report commissioned by the European Travel Research Foundation, London.

Thomas, M. (1987) 'Customer care: the ultimate marketing tool', in Wensley, R. (ed.), *Reviewing Effective Research and Good Practice in Marketing*, Proceedings of the Marketing Education Group, Warwick University, pp. 283–294.

Tordjman, A. (1994) 'European retailing: convergences, differences and perspectives', *International Journal of Retail and Distribution Management*, 22 (5), pp. 3–19.

Wachter, M. (1974) 'Primary and secondary labour markets: a critique of the trial approach', *Brookings Papers on Economic Activity*, No. 3, pp. 637–693.

Walsh, S. (1992) 'Airport shopping centre design', *AACI Airport Conference*, Vienna.

Walters, D. (1988) *Strategic Retailing Management: A Case Study Approach*, Prentice-Hall, Hemel Hempstead, UK.

Walters, D. and White, D. (1987) *Retail Marketing Management*, Macmillan, London.

Weber, P. (1995) 'Innovative commercial strategies', unpublished mimeo, BAA Retail Services, London.

Wilkinson, F. (1981) *The Dynamics of Labour Market Segmentation*, Academic Press, London.

Williamson, O.E. (1975) *Markets and Hierarchies Analysis and Anti Trust Implications*, Free Press, New York.

Wrigley, N. (1993) 'Retail concentration and the internationalisation of British grocery retailing', in Bromley, R. and Thomas, C. (eds), *Retail Change: Contemporary Issues*, UCL Press, London, pp. 41–68.

Wrigley, N. (1994) 'After the store wars: towards a new age of competition in UK food retailing?', *Journal of Retailing and Consumer Services*, 1, pp. 5–20.

Wrigley, N. (1996) 'Sunk costs and corporate restructuring; British food retailing and the property crisis', in Wrigley, N. and Lowe, M. (eds), *Retailing Consumption and Capital: Towards the New Retail Geography*, Longman, London.

WTS (1995) *World of Travel Shopping*, Generation AB Publications, Ornskoldsvik, Sweden.

WTTC (1995) *Economic Contribution of Travel & Tourism Enormous for the Global Economy*, World Travel and Tourism Press Release, London.

Index